A Fever in Salem

A Fever in Salem

A *New Interpretation of the*
New England Witch Trials

LAURIE WINN CARLSON

CHICAGO *Ivan R. Dee* 1999

Library of Congress Cataloging-in-Publication Data:
Carlson, Laurie M., 1952–
  A fever in Salem : a new interpretation of the New England witch trials / Laurie Winn Carlson.
      p. cm.
  Includes bibliographical references and index.
  ISBN 1-56663-253-6 (alk. paper)
    1. Witchcraft—Massachusetts—Salem—History—17th century.
  2. Epidemic encephalitis—Massachusetts—Salem—History—17th century. I. Title.
  BF1576.C37 1999
  133.4'3'097445—dc21   99-27520

To Molly

ACKNOWLEDGMENTS

Many people have helped me see this project along its way, particularly Dr. Mike Green of Eastern Washington University, who never lost enthusiasm for the project. Oliver Sacks, M.D., was generous with useful advice; his work was most inspiring. Thanks as well to Robert Shope, M.D., of the University of Texas Medical Branch for valuable suggestions. I am grateful to the following, who gave me research tips, pointed me in the right direction, or shared valuable advice: Dr. William H. Calvin, University of Washington; Prof. John S. Oxford, Royal London Hospital; Thomas Rosenbaum, archivist, Rockefeller Foundation; Michelle Perry, Parkinson's Disease Society, London; Simon Warburton and Lesley A. Hall, Wellcome Institute for the History of Medicine; Chester Moore, Division of Vector-Borne Infectious Diseases, Centers for Disease Control, Fort Collins, Colorado; Jane E. Ward, Peabody Essex Museum, Salem; Will La Moy, Phillips Library, Peabody Essex Museum, Salem; Elizabeth Tunis, History of Medicine Division, National Library of Medicine. At Eastern Washington University I found the advice of Kathleen Huttenmaier, Dr. Susan Stearns, Dr. Ann Le Bar, Dr. Laura Phillips, Janelle Braithwaite, and especially Joy Scott very helpful. I very much appreciate the assistance of my editor, Hilary Schaefer, and my publisher, Ivan R. Dee. Thank you as well to my husband, Terry, a partner and friend.

The desire to find the significant detail plus the readiness to open his mind to it and let it report to him are half the historian's equipment. The other half [is] the idea, point of view, the reason for writing. . . . The art of writing is the third half. If that list does not add up, it is because history is human behavior, not arithmetic.

—Barbara Tuchman, *Practicing History*

Science is an attempt, largely successful, to understand the world, to get a grip on things, to get hold of ourselves, to steer a safe course. Microbiology and meteorology now explain what only a few centuries ago was considered sufficient cause to burn women to death.

—Carl Sagan, *The Demon-Haunted World*

CONTENTS

PREFACE

Such was the darkness of that day, the tortures
and lamentations of the afflicted, and the power
of former precedents, that we walked in the
clouds and could not see our way. . . .
   —Reverend John Hale of Beverly Farms,
   A Modest Inquiry, 1697[1]

During the latter part of the seventeenth century, residents
of a northeastern Massachusetts colony experienced a suc-
cession of witchcraft accusations resulting in hearings, trials,
imprisonments, and executions. Between 1689 and 1700 the
citizens complained of symptoms that included fits (convulsions), spectral visions (hallucinations), mental "distraction"
(psychosis), "pinching, pin pricking and bites" on their skin
(clonus), lethargy, and even death. They "barked like dogs,"
were unable to walk, and had their arms and legs "nearly
twisted out of joint."

  In late winter and early spring of 1692, residents of
Salem Village, Massachusetts, a thinly settled town of six
hundred,[2] began to suffer from a strange physical and mental malady. Fits, hallucinations, temporary paralysis, and
"distracted" rampages were suddenly occurring sporadically
in the community. The livestock, too, seemed to suffer from
the unexplainable illness. The randomness of the victims

and the unusual symptoms that were seldom exactly the same, led the residents to suspect an otherworldly menace. With the limited scientific and medical knowledge of the time, physicians who were consulted could only offer witchcraft as an explanation.

These New Englanders were Puritans, people who had come to North America to establish a utopian vision of community based upon religious ideals. But, as the historian Daniel Boorstin points out, their religious beliefs were countered by their reliance on English common law. The Puritans did not create a society out of their religious dogma but maintained the rule of law brought from their homeland. They were pragmatic, attempting to adapt practices brought from England rather than reinventing their own as it suited them. When problems arose that were within the realm of the legal system, the community acted appropriately, seeking redress for wrongs within the courts.[3]

Thus when purported witchcraft appeared, church leaders, physicians, and a panicked citizenry turned the problem over to the civil authorities. Witchcraft was a capital crime in all the colonies, and whoever was to blame for it had to be ferreted out and made to stop. Because no one could halt the outbreak of illness, for ten months the community wrestled with sickness, sin, and the criminal act of witchcraft. By September 1692, nineteen convicted witches had been hanged and more than a hundred people sat in prison awaiting sentencing when the trials at last faded.[4] The next year all were released and the court closed. The craze ended as abruptly as it began.

Or did it? There had been similar sporadic physical complaints blamed on witchcraft going back several decades in New England, to the 1640s when the first executions for the crime of witchcraft were ordered in the colonies.[5] Evidence indicates that people (and domestic animals) had suffered similar physical symptoms and ailments in Europe in still earlier years.[6] After the witch trials ended in Salem, there continued to be complaints of the "Salem symptoms" in Connecticut and New Hampshire, as well as in Boston, into the early eighteenth century. But there were no more hangings. The epidemic and witchcraft had parted ways.

By examining the primary records left by those who suffered from the unexplainable and supposedly diabolical ailments in 1692, we get a clear picture of exactly what they were experiencing. *The Salem Witchcraft Papers*, a three-volume set compiled from the original documents and preserved as typescripts by the Works Progress Administration in the 1930s, has been edited for today's reader by Paul Boyer and Stephen Nissenbaum. It is invaluable for reading the complete and detailed problems people were dealing with. Like sitting in the physician's office with them, we read where the pain started, how it disappeared or progressed, how long they endured it.

A similar epidemic with nearly exact symptoms swept the world from 1916 to 1930. This world-wide pandemic, sleeping sickness, or encephalitis lethargica, eventually claimed more than five million victims. Its cause has never been fully identified. There is no cure. Victims of the twentieth-century epidemic continue under hospitalization to the present day. An excellent source for better understanding encephalitis lethargica is Oliver Sacks's book *Awakenings*, which is now in its sixth edition and has become a cult classic. A movie of the same title, based on the book, presents a very credible look at the physical behaviors patients exhibited during the epidemic. While encephalitis lethargica, in the epidemic form in which it appeared in the early twentieth century, is not active today, outbreaks of insect-borne encephalitis do appear infrequently throughout the country; recent outbreaks of mosquito-borne encephalitis have nearly brought Walt Disney World in Florida to a halt, have caused entire towns to abandon evening football games, and have made horse owners anxious throughout the San Joaquin Valley in California.[7]

Using the legal documents from the Salem witch trials of 1692, as well as contemporary accounts of earlier incidents in the surrounding area, we can identify the "afflictions" that the colonists experienced and that led to the accusations of witchcraft. By comparing the symptoms reported by seventeenth-century colonists with those of patients affected by the encephalitis lethargica epidemic of the early twentieth century, a pattern of symptoms emerges. This pattern supports the hypothesis that the witch-hunts of New England were a response to unexplained physical and neurological behaviors resulting from an epidemic of encephalitis. This was some form of the same encephalitis epidemic that became pandemic in the 1920s. In fact it is difficult to find anything in the record at Salem that *doesn't* support the idea that the symptoms were caused by that very disease.

A Fever in Salem

CHAPTER ONE

The Witch Craze in Seventeenth-century New England

/\\

'Tis now the very witching time of night,
When churchyards yawn and hell itself breathes
    out
Contagion to this world: now could I drink hot
    blood,
And do such bitter business as the day
Would quake to look on.
                  —Shakespeare (1564–1616), *Hamlet*

John Hale, a pastor in Beverly, Massachusetts, a town just
north of Salem, in 1702 wrote an overview of the events related to the witch craze in New England. He resided in the
area and knew many of the Salem Village residents, as several of them had attended his church in Beverly before the
village established its own church. Satan had raised much
trouble, Hale observed, "the beginning of which was very
small, and looked on at first as an ordinary case." From the
first one or two persons examined for suspected witchcraft,
the problem expanded until "a multitude of other persons
both in this and other Neighbour Towns, were Accused, Ex-

amined, Imprisoned, and came to their Trials, at Salem, the County Town, where about Twenty of them Suffered as Witches; and many others in danger of the same Tragical End: and still the number of the Accused increased unto many Scores." Many of the accused, Hale pointed out, were persons of "unquestionable Credit, never under any grounds of suspicion of that or any other Scandalous Evil."

Not everyone was comfortable with the events as they unfolded: "this brought a general Consternation upon all sorts of People, doubting what would be the issue of such a dreadful Judgment of God upon the Country." Their fears subsided when the accusations diminished in the autumn of 1692. Still, bitterness and fear lingered: "it left in the minds of men a sad remembrance of that sorrowful time; and a Doubt whether some Innocent Persons might not Suffer, and some guilty Persons Escape." Twenty had been hanged and dozens of others released only after confessing to witchcraft. Perhaps, Hale wrote, there had been many mistakes made by the judges and juries in their zeal to punish sin. He reasoned with hindsight that the "Laws, Customs, and Principles used by the Judges and Juries in the Trials of Witches in England," which the colonists had used as "patterns" in New England, may have been "insufficient and unsafe."[1] At the time, witchcraft was a capital crime in every one of the New England colonies, just as it was in England.

The ideology of witchcraft had evolved from beginnings as an ancient fertility cult. It was modified in the sixteenth century when agrarian peasant societies assimilated a variety of popular beliefs into an increasingly diabolical witchcraft culture. As earlier folk beliefs changed, benevolent witches were replaced by evil entities. Since the Enlightenment, scholars have rarely been interested in the afflicted but rather in the witches and their confessions, and they have concentrated on the barbarity and irrationality of the prosecutions for witchcraft. Yet few studies until the late nineteenth century connected witches' confessions to hallucinatory drugs or pathological states, particularly hysteria.[2] In the mid-twentieth century a revival of interest in witchcraft prompted a new flood of interpretations of witch trials.

Because outbreaks of witchcraft were centered in rural agrarian communities, in both the North American colonies and in western Europe, settings where collective cooperation was imperative for survival, these communities were vulnerable in a way that more densely settled urban areas were not. Stresses related to warfare, both the Thirty Years' War in Europe and the French and Indian Wars of New England in the latter 1600s, have also been thought to have played a part in the appearance of witches. In small communities where residents relied on one another, everyone's fate was intertwined, and if someone within the community—a friend or neighbor—had the ability and motive to cause affliction and death, the horror was intensified. Whom could one trust?

Historically witches are not the giggly Samanthas of 1960s television shows or the cartoonish characters we see in today's Halloween decor. Through the centuries witchcraft has meant horror and violence, torture and execution of innocents, and an unresolved fear of the unknown. In seventeenth-century New England villages those fears were not unfounded: chubby "thriving" babies were in the throes of heart-rending convulsions; suddenly blinded farmers were unable to work (in an era with no social safety net); precious, well-behaved children were suddenly uncontrollable; and nearly everyone knew someone who had gone mad from a "distemper" and died inexplicably. These losses were harsh reality, but what made them more difficult to accept was the absence of explanation for the "afflictions." Without one, the reason came to be evil in the form of the Devil. A lurking, sinister, unexplainable presence had to be to blame. And how could anyone know exactly where or how Satan might appear? One's neighbor, spouse, parent— anyone might be culpable. Fear ran rampant through New England, and in 1692 events moved quickly beyond anyone's control.

Besides John Hale, many other disgusted and disapproving residents of the Salem region knew too well that the witch-hunts, trials, and executions had gone too far. Looking back at the events after the crisis period, people were more likely to regret the brutal treatment and hangings. But only months earlier, a different atmosphere had permeated New England towns and villages. A fierce, implacable foe appeared to have been loose, and the residents seemed helpless to understand or end the rampant incidents.

Salem's witch trials are etched in American history almost as folklore; yet the persecution of witches in America extended over time, from the 1640s to 1692, and geographically beyond Salem to include surrounding New England towns. During this period something unexplainable and distinct from known illness caused people *and* domestic animals to behave strangely. This unseen force caused people to fall into fits, feel pains in their arms and legs like biting and pricking, bark like dogs, grovel on the ground like hogs, and even turn suicidal. Psychotic hallucinations were frightening enough, but when an individual's eyes twisted to the side, and arms and legs stiffened in awkward postures for hours, or blood wept from marks on the skin that clearly looked like bites, everyone knew that something was terribly wrong. Even the colonists' animals were struck with unusual symptoms and sudden death at the same time the villagers were suffering.

In reaction to these behaviors, the community first turned to its educated leaders: physicians and ministers. Medical practitioners—"doctors of physick"—tried various remedies but could offer no explanation beyond that the afflictions were "otherworldly." Ministers were called in to determine a spiritual course of action, which included communal prayers and fasting.

In a community based on law, the Puritans turned next to their court system to resolve the situation. Hearings, held publicly and recorded, brought everyone involved into the public eye. Accused witches were questioned; afflicted persons were brought forward to tell who caused their pain; family members testified about those who might wish to harm them or others. Magistrates listened, questioned, and dismissed or convicted. One by one, those individuals found guilty of causing the problems—guilty of the crime of witchcraft—were hanged by the neck.

But executions did not still the outbreak of complaints. Throughout the summer of 1692, more and more people came forward to accuse and testify against others. In Salem the hanging of witches began to resemble human sacrifices to an angry god who would not halt the fits, convulsions, and terror. Twenty accused witches were executed, and four died in prison that summer. By October more than a hundred still lingered in jail, but by then the siege was over. In the last weeks of October, as a hard frost hit New England gardens and ice etched the rims of ponds, a group of ministers led by Increase Mather petitioned Massachusetts Governor Phipps to close the court. By the spring of 1693 all the accused had been released; the Salem witch trials were over. It was the last of the New England witch-hunts; the horror and strife seemed to end as quickly as it had begun.

CHAPTER TWO

The Afflicted

And Ever Since this Child hath been followed
with grevious fitts as if he would never recover
more; his head & Eyes drawne aside So as if they
would never Come to rights . . . lying as if he
were in a manner dead, falling anywhere Either
into fire or water if he be not Constantly looked
to, and generally in Such an uneasie and restless
frame allmost allways running too & fro acting so
Strange that I cannot judge otherwise but that he
is bewitched and by these circumstances do believe that the aforesaid Bridget Oliver is the
Cause of it and it has been the Judgment of Doctors Such as lived here & foreigners; that he is
under an Evil hand of Witchcraft.
> —Samuel Shattuck and Sarah Shattuck,
> June 2, 1692[1]

Most narratives of what happened at Salem begin with the
household of Reverend Samuel Parris, pastor of the Salem
Village Church, whose nine-year-old daughter and eleven
year-old niece began experiencing odd behavior in midwinter 1692. Reverend John Hale, of nearby Beverly, described
how the girls had been afflicted by a strange "Distemper"
which Parris did not recognize. He "made application to

Physitians," but the strange behavior grew worse. Eventually one puzzled doctor suggested that they were "under an Evil Hand," and that medicine could do nothing to relieve the suffering girls. Neighbors by that time had become involved, as was common in close-knit New England communities, and quickly concluded that the girls were bewitched. By this time the girls were crying out, claiming to be pinched, pricked with pins, and "grievously tormented" with visions that no one else could see. Hale writes that these "Children were bitten and pinched by invisible agents; their arms, necks, and backs turned this way and that way, and returned back again, so as it was impossible for them to do of themselves, and beyond the power of any Epileptick Fits, or natural Disease to effect." The girls were sometimes unable to speak or hear, "their throats choaked, their limbs wracked and tormented so as to move an heart of stone to sympathize with them. . . ." The neighbors were filled with compassion and concern for some way to alleviate their "cruel Sufferings" as well as to explain the cause. Reverend Parris called in "some Worthy Gentlemen of Salem" as well as some "Neighbor Ministers" to consult at his house. After meeting and examining the girls, all agreed that the problem was "preternatural" and they feared that "the hand of Satan was in them."[2]

The residents were, after all, not unaware of other similar cases. The girls were "in all things afflicted as bad as John Goodwins children at Boston, in the year 1689." In order to understand what was affecting the Parris household, and why the father as well as neighbors were so concerned, we must examine the case of the Goodwin children as chronicled by Cotton Mather in 1689, in his book titled *Memorable Providences Relating to Witchcrafts and Possessions.* Mather, a significant figure in the Salem events, was a prolific writer; he was continually working on a new book. When he was called in to examine the children of the pious Goodwin family in Boston, people who belonged to his own church, he immediately realized the significance of what he was seeing and recorded extensive notes about his observances. Mather was trained in both medicine and the ministry; he had such a problem with stuttering that he had studied medicine thinking he was unsuited for a career in the ministry. But he was able to overcome his speaking problem and became a member of the Boston clergy, following the lead of his illustrious father, Increase Mather. Cotton combined the thinking of an Enlightenment scientific reader with the dogma of Puritan Christianity. Historians have been critical of Cotton Mather's ability as well as his character, but he was an ordinary man living with one foot in two different eras. His observance and notation of the children's behavior indicate that he had a grasp of scientific thinking, but once he recognized that something was unusual, he did not have the medical knowledge to understand it. He fell back on the only thing he trusted: religion.

The four afflicted Goodwin children (the eldest thirteen, the youngest five) had experienced the ordinary family life provided by hardworking, devoutly religious parents. Their father, John Goodwin, was a mason. In midsummer 1688 the thirteen-year-old daughter, Martha, was "visited with strange Fits, beyond those that attend an Epilepsy, or a Catalepsy, or those that they call The Diseases of Astonishment," Mather noted. Shortly after, one sister and two brothers (two other Goodwin children were never affected) "were seized . . . with affects like those that molested her." Within a few weeks the children were "tortured everywhere in a manner so very grievous, that it would have broke an heart of stone to have seen their Agonies."[3]

The Goodwins contacted physicians first. Dr. Thomas Oakes, a prominent physician who later became speaker of the legislature and a colonial deputy to England, was so puzzled by the children's ailments that he felt "nothing but an hellish Witchcraft could be the origin of these Maladies." The children's "tortures" increased; "sometimes they would be Deaf, sometimes Dumb, and sometimes Blind, and often all this at once." No one knew what to think: "one while their Tongues would be drawn down their Throats; another while they would be pull'd out upon their Chins, to a prodigious length. They would have their Mouths opened unto such a Wideness, that their Jaws went out of joint; and anon they would clap together again with a Force like that of a strong Spring-Lock. The same would happen to their Shoulder-Blades and their Elbows, and Hand-wrists, and several of their joints." The children were experiencing extremely violent seizures which had never before been seen in Boston, or at least had not been seen by the learned men of the town. When the seizures subsided, the children would lie "in a benumbed condition," likely in a comalike state. Other times the children's bodies would be stretched so awkwardly that witnesses "fear'd the very skin of their Bellies would have crack'd."[4] The Goodwin household was anything but calm.

While "scores" of onlookers tried to discern whether the children's problems were physical or spiritual, the children "would make most pitteous outcries, that they were cut with Knives and struck with Blows that they could not bear. Their Necks would be broken, so that their Neck-bone would seem dissolved unto them that felt after it; and yet on a sudden, it would become again so stiff that there was no stirring of their Heads; yea, their Heads would be twisted almost round." The children lay still between fits, "most pitiful spectacles," and the suffering went on for weeks. Mather tried praying with one child, but to no avail—"the Child utterly lost her Hearing till our Prayer was over."[5]

In Boston the children would have been able to avail themselves of the highest-educated physicians, men schooled in Britain or Europe. But the Goodwin parents were a "Religious Family" and determined to cure their children, "to oppose Devils with no other weapons but Prayers and Tears. . . ." The weaponry available to medical men at the time was a wide array of medications that were basically "simples" based on herbals. The sixteenth-century pharmacopoeia consisted of St. John's wort, Clown's All-Heal, spurge, fennel, saffron, parsley, elder, snake-root, opium in some form, roasted rhubarb, jalap, black hellebore, and occasionally iron and antimony. At least some of these remedies were probably tried but were not strong enough to alleviate the Goodwin children's misery. After the physicians admitted they could do nothing, the Goodwins requested that the four ministers of Boston and one from Charlestown perform a Day of Prayer at their house. The youngest child became "miraculously cured" that very day.[6]

The child's cure, linked to prayer, solidified the idea that there was something supernatural about the children's problems. No longer did anyone consider them to be physically based. Enter the magistrates, who sought to punish whomever had caused the "possessions." Ultimately poor Mrs. Glover, an Irish washerwoman Mather referred to as "the Hag," was tried for the crime. She was a Roman Catholic, and when tested by the magistrates was found unable to recite the Lord's Prayer correctly (Mather noted that she could say it in Latin, "very readily," but that did not count). She was examined by the physicians, found to be of sound mind, and was sentenced to prison.[7] Salem's witchcraft era began with the Goodwin children and Mrs. Glover. From that point, all similar physical afflictions were viewed as caused by Satan, and could be alleviated only by prayer (rather than medicine) or punished by law.

But the Goodwins were not the first strange case nor the only one at the time in Boston. While Mrs. Glover was being examined in court, one witness told of a woman who had been "bewitched to death" after she had experienced hallucinations. The witness's own young son had broken into fits "in the same woful and surprising manner that Goodwins children were." The boy's affliction passed and he recovered—which people attributed to the timely jailing of Mrs. Glover.[8]

The Hag's unfortunate imprisonment, however, did not relieve the other three Goodwin children, who still "continued in their Furnace as before, and it grew rather seven times hotter than it was." All their afflictions persisted, with the addition of new ones: they barked "like Dogs" and purred "like so many Cats." They sweated, and panted, and cried out that someone was hitting them. Onlookers drenched the children in cold water but to no avail. The ministers saw no one hit the children but were dumbfounded to see "the Marks left by them in Red Streaks upon their bodies afterward."[9]

The children screamed that they were being roasted on a spit or being slashed by knives; they would either be "so Limber, that it was judg'd every Bone of them could be bent," or their muscles would tense so that they were completely stiff. They would sometimes be "as though they were mad," and their frenzied running, leaping, and "flying" were "beyond the Imagination of them that look'd after them."[10]

These innocent, God-fearing children began to be dangerous: they tried to harm others by striking and hitting, or trying to push their "tenderest and dearest friends" down the stairs. They tried to harm themselves, very near burning in the open fireplace or drowning. One almost choked by pulling his own neck clothes tight around his throat, being stopped from strangling by a caretaker.[11]

Due to the children's physical debilities, it took "an Hour or Two" to dress them; they were so excessively overactive and out of control that it was hard to keep bedclothes on the bed or the children. Their muscular contortions made everything difficult—even hand-washing. Their hands were clasped together so rigidly that there was no way to relax their frozen postures. Eating meals was difficult because their jaws were clamped shut and no food could be put into their mouths.[12]

Mather, seeing an opportunity to gain material for the book he was writing, took Martha, the eldest, home in order to study her at close range. She was industrious and "pious," but in mid-November she was overtaken by paranoia, saying, "Ah, *They* have found me out!" and her fits immediately resumed. In seventeenth-century parlance, Mather describes how she coughed up a "Ball as big as a small Egg" into her throat, where it settled on the side of her windpipe. She was nearly choked by it before "Stroking and Drinking" allowed her to swallow it again. A swollen gland in the neck was evidently something that Mather, educated in the medicine of the day, had never before encountered.[13]

Martha Goodwin suffered with her teeth set, her eyes receding into her head until Mather "fear'd she should never have used them" again. She experienced loud ringing noises in her ears. Her senseless, "ludicrous Fits" continued; she threw herself on the floor and tried to jump to the bottom of the well because she insisted a plate needed retrieving.

When calm returned, her eyes would be "strangely twisted and blinded," and her neck distorted.[14]

The ministers held prayer days with the family, and eventually the children all grew well—which was attributed to prayer. Mather reported that Martha finally recovered, having only a slight twisting or tic of her eye muscles and a cough. The other Goodwin children's fits abated, and though at first they were "always quiet," they were shortly able to work, read,* and resume normal activities.

Pneumatology, the scientific study of angels and demons, made complete sense to Cotton Mather. He worked on a book, eager to use it as proof that devils really did exist as well as witches. After the ordeal of the Goodwin children, he declared, anyone who doubted Satan or witchcraft was both ignorant and dishonest. Who could (or would) argue with him?[15]

The Goodwin children became the benchmark for the outbreaks that followed. Less than four years later, in 1692, residents of the region were wracked with the same frightening afflictions. Besides the two girls in the Parris household at Salem Village, others began to experience odd and unusual fits and "afflictions": Mary Walcott, the seventeen-year-old daughter of the commander of the Salem Village militia; Elizabeth Hubbard, another seventeen-year-old, the niece and maidservant of Dr. Griggs, the town's physician; Mrs. Pope, a "woman of good social position and in early middle life"; and Sarah Vibber, a thirty-six-year-old married woman.[16]

Although the Goodwin children and these other cases

---

* The ability to read was a test to examine cognitive abilities. Mather frequently shoved open books before people in their fits, in order to gauge their condition.

were most immediate to mind, there had been earlier similar cases in New England. In 1672 Reverend Samuel Willard had written extensive observations of Elizabeth Knapp, a sixteen-year-old girl who had exhibited "strange and unusual" seizures the previous year. Her shrieks of pain, bursts of laughter, and falling to the ground had puzzled him. He figured she was ill, but when he asked her if she felt sick she insisted that she did not. It made him "wonder." Knapp had pains in the legs, chest, throat, a choking sensation, exhibited "roarings and screamings," and persisted in leaping and running around.[17]

Her fits had begun October 30, 1671, occurred intermittently, and returned in mid-December. She attempted to kill herself and was violent to others. Her eyes were "sealed up" during the fits, and she was "speechless." She sometimes cried out nonsensically, such as repeating the words "money, money," or she "barked like a dog and bleated like a calf." She was coherent afterward and could recite everything that had been said to her during the episode. Her fits alternated with quiet periods when she apologized and prayed for forgiveness for her sinful behavior.[18]

Her possession spanned three months; some periods of fits were sixteen to eighteen days in duration. On January 15 the record ends; we do not know what happened to her.[19]

In addition to Elizabeth Knapp, there were other cases before 1692. Another incident in Newbury, Massachusetts, began in November 1679. William and Elizabeth Morse's grandson, John Stiles, had come to live with them. The boy developed "fits, afflictions and hurried into great motions" or was "flung about in such a manner as they feared that his Brains would have been beaten out." He claimed of being pricked and pinched and tried to throw himself into the fire.

He was "miserable lame," then would rave and bark like a dog and even tried to eat sticks, ashes, and rug-yarn.[20]

Like Elizabeth Knapp, young John Stiles had periods of respite and calm; but the fits always returned. Eventually Elizabeth Morse, a midwife, was accused of bewitching her own grandson, found guilty, and sentenced to death by hanging. Due to a delay, she remained imprisoned until the following spring when she was allowed to return to her home under a form of house arrest. Her execution was not pursued; her husband died two years later. There are no records of John Stiles after that.[21]

The first New England witch trial, which resulted in the hanging of Alice Young, had been held in Windsor, Connecticut, in the spring of 1647. It accompanied a severe "epidemical sickness" which killed many that year. In Massachusetts, too, the first full-fledged witch trials coincided with epidemics. This period coincides with a peak in the intense witch-hunt frenzy in 1645 in Europe.[22]

Even though such intermittent "distractions" and "bewitchments" had occurred in New England, events intensified in the Salem area in 1692 because of the high death rate due to fits. No longer was it a matter of a housewife or two "bedrid" or made "insensible" for years. In 1690 Priscilla Stacy, a little girl, had died after two weeks of unusual behavior during which she "screaked out" during violent fits. Salem residents still remembered Thomas Green's twins who died of fits sometime between 1690 and 1692. James Beale's boy, of Marblehead, died in March 1692. Mary Warren's mother had "taken ill and dyed" in an odd manner too. There had been three well-known deaths in 1689 in the Salem area: adults John Fuller and Rebecca Shepherd died strangely violent deaths from a "malignant fever." Two days later Ben Holton died from violent fits exactly like theirs.[23]

When the little Putnam girl, Ann, became fitful in early spring 1692, her parents and grandparents were frantic—Rebecca Shepherd had been her aunt. Their worries were increased by the fact that the father's brother, the town marshal, became sick at the same time with fits and hallucinations (but recovered) while his two-month-old infant died from similar symptoms.[24] Grief and fear permeated Massachusetts; people seemed to be succumbing to a strange and powerful ailment.

Samuel Shattuck's child was a well-known case in Salem. The child had begun having fits in 1680 and twelve years later was still in a catatonic state. Samuel explained how the boy, their "Eldest Child" who had been in good health, was taken in a "very drooping Condition" and began behaving "in a strange and unuseall maner," as if his insides would burst from his chest. The boy lay so strangely that the "Doctor & others did beleive he was bewitched." Sometimes he would stand at the door and fall to the ground, bruising his face on a stepping stone as if he had been "thrust out bye an invisible hand." He fell frequently, "in a very miserable maner." The sorrowful father detailed the boy's fits: "his mouth & Eyes drawne aside and gasped in Such a maner as if he was upon the point of death." The fits increased in intensity, after which he would spend "many months" crying until he was exhausted.

Eventually the parents realized their worst fears: "wee p'rceived his understanding decayed Soe that wee feared . . . that he would be quite bereaft of his witts; for Ever Since he has bin Stupified and voide of reason." The child, his mind gone, continued experiencing the fits. His mother had tried to get him to walk by laying a board flat on the ground for him to step on. He would go only so far, then would halt, unable to continue. She tried everything—coaxing, luring

(with "Cake & mony"), begging—but he could not go past
the end of the board without help. She tried it with him several times, but on any board he could not walk past the end.
The parents figured "some inchantm't" kept him from
going farther. His behavior was similar to that of other afflicted children: he would have fits with his head and eyes
drawn aside, his neck twisted beyond endurance, falling either into the fire or the water "if he be not Constantly
looked too." He was generally in "Such an uneasie and rest-
les frame allmost allways running too & fro acting soe
Strange" that everyone judged that it had to be witchcraft. It
had been difficult for the Shattucks to accept. They testified at two trials on behalf of their son, accusing Bridget
Bishop as well as Mary Parker of causing his problems.
Samuel Shattuck said they were not alone in their judgment, as "it has bin the Judgem't of Docters Such as lived
here & forreigners: that he is under an Evill hand of witchcraft."[25]

Some of the Shattucks' neighbors had tried to be helpful
by cutting pieces of the boy's hair and frying it in a skillet in
the fireplace, then tossing it on the floor—a folk remedy for
identifying evil spirits. The next person to enter the house,
then, was believed to be responsible for bewitching the poor
child. Mary Parker unknowingly came over to see if the
Shattucks were interested in purchasing some chickens from
her and became the putative witch who in September 1692
was tried and executed. The boy never recovered; he continued in "a very Sad condition" with fits that had "taken
away his understanding."[26]

The unfortunate suspect, Mary Parker, had been sick
herself. The previous January (1692) she had been found
lying on the dirt and snow in a comalike state. At her witch
hearing, others agreed that it was not the first time they had

seen her in "such kind of fits," as she was known to have had frequent episodes. At the time, neighbors had carried her indoors, fearing she was dead, but after they put her in a warm bed she suddenly burst up in a fit of hysterical laughter. As such behavior was incomprehensible, their sympathy for her evaporated.[27]

In March of that year, the Fuller family's maidservant, Betty, was found lying outdoors on the ground. Fearing she had fallen "Downe Ded," some young men took her indoors and after three hours without "any motission of Life" she was able to speak and said she had seen a vision of a "womman with A white cape."[28]

The Perleys were another couple with a sad family tale: their ten-year-old daughter had been in a "soroful condition," complaining of being "Pricked with Pins" and sometimes falling down into "dredful fits" or trying to throw herself into the open fireplace or well. They consulted several doctors, but they "tould us that she was under an evil hand." There was nothing the parents or doctors could do for her; she was sick for three years and wasted away to "skin and bone," finally ending "her sorowful life." Ruth Perley, the dead girl's mother, signed the court document with her mark. Her father continued the deposition in court at a witch trial where the couple accused Elizabeth How of their daughter's troubles.[29] How was executed on July 19, 1692.

John Hale of Beverly, in a deposition at Bridget Bishop's witch trial, described the afflictions that had befallen his friend's wife five or six years earlier. He told how Christian Trask, living with her family on the edge of Salem where it bordered the town of Beverly, had begun to act strangely: she "was distracted" so much that her family had sought help through prayer and fasting. She seemed to pull herself together, and Hale "took it then to bee only distraction, yet fearing sometimes somw't worse"; but now that he had been viewing events in Salem, he had "seen the fitts of those bewitched at Salem Village call to mind some of hers to be much like some of theirs." The story has a sad ending: Christian's condition worsened; erupting into fits at public meetings and needing to be carried home, she was "sometimes better sometimes worse," sometimes suicidal and attempting to hurt others. Hale described her eventual death by suicide—she had torn her throat and jugular vein with a pair of scissors. He could not believe she had done it to herself and insisted that it must have been due to "some extraordinary work of the devill or witchcraft."[30]

Phoebe Chandler, twelve years old, related her experience: "about one half of my right hand was greatly swolen & exceeding painful, & allsoe part of my face," which of course she could not account for. It continued very badly along with a great weight on her chest and legs, so much so that she could hardly walk. She had intense burning in her stomach and was struck deaf. Hers was a mild sort of affliction, exceedingly common in New England at that time. Several people reported heavy weights on their chests; John Louder had experienced it back in 1684, so had Jarvis Ring of Salisbury. They were young men at the time. In 1692 Sarah Vibber, thirty-six years old, and Ann Putnam, thirty years old, also reported it, along with John Parker's son, age not known.[31]

Most court depositions were made by witnesses on behalf of the afflicted individuals, who were frequently unable to talk (being in a catatonic state, or "deaf and dumb"), if indeed they were still alive. It was a neighborly thing to do, and besides, the afflicted suffered so horrendously that other citizens felt responsible for trying to put an end to their suffering. George Herrick and John Putnam both saw Mercy Lewis "in a very Dreadfull and Solemn Condition: so that to our apprehention She could not continue long in this world" unless something was done to "mitigate those torments." That desire to save her caused them to "Expediate a hasty dispatch" to apprehend a suspected witch, Mary Easty, "in hopes of possable it might save her Life." It was definitely a twisted form of citizenship and caring, but the men felt they were motivated by purely selfless reasons to ferret out the witch, bring her to face what she had done, and make amends. They were not motivated by vengeance as much as by their desire to save the community from further harm.

Two other middle-aged men related how moved they were by the way the young women were "most dreadfully tormented," being "torne all to peaces or all their bones putt out of joynt"; the men reported having seen "dreadful marks in their fleesh" that were "such tortors as no tongue can express." If witches were to blame, the men of the village were not about to sit back and let them wreak havoc on the community. They came forward to swear in court that the girls had mentioned who was causing their torments— and the hearings and trials continued.[32]

Along with the fits that were so fully described in the court documents, there were many references to other physical concerns, for instance the pricking by pins and biting, evidenced by painful marks on the afflicted individual's skin. Many people witnessed them and even watched as the marks mysteriously appeared on a victim's skin while the person was bedridden or in court. Abigail Williams cried out in pain during one trial, claiming that she was being pinched. Onlookers saw "great printes" on her arm. Yet no one had seen anyone suspicious near her. Little Ann Putnam, suffering from fits and pain, had been "bitten" by a witch, her uncle testified. He knew the exact time, about "2 of the clok" when he saw the strange marks, "the mark being in a kind of a round ring and 3 stroaks a Cros the ring," which he supposed looked like she had been struck with a chain brandished by a witch. He watched as she received "6 blows with a Chane in the space of half an ouer and she had one remarkable one with 6 stroaks a Cros her arme." He testified under oath and signed his name that he "saw the mark boath of bite and Chane." Mary Warren was another young woman whose "bites had been seen by many."[33]

The victims were not only young females. Jarvis Ring, from nearby Salisbury, swore an oath that he had had the same problems, from about 1684 to 1685, when he had experienced paralysis, inability to speak, hallucinations, and bites and pinching. One participant in the events has been difficult for historians to assess: John Indian, a West Indian slave who belonged to Reverend Parris, also fell into fits in court and at a public tavern. Unlike the pathetic little girls and young maids, Indian did not win much support then, or from historians today, but he too fell to the ground in agonizing writhing fits and cried out that he was bitten. When she was summoned to court to defend herself against accusations of witchcraft, brash Sarah Good boasted that she was not worried, there was only one "Evidence" and "thats an Indian." She carried on a tirade against the magistrates, however, and witnesses claimed she tried to kill herself before trial.[34]

Bruises, blood drawn to the surface of the skin, these "marks of plaine bites on there flesh" looked so like teeth marks that the alarmed witnesses were more intent on finding out who, rather than what, was to blame. Ailments had taken a very personal dimension—finally a visible mark could be seen on the skin. The dental hygiene of the English colonists was terrible; few of the colonists had any teeth, particularly adult women, who were described in 1674 by a traveler: "the Women are pittifully Tooth-shaken; whether from the coldness of the climate or by sweetmeats of which they have store, I am not able to affirm." The treatment for toothache at the time was sulfur and gunpowder, mixed to a paste. Many of those accused of witchcraft—and of biting—could not possibly have accomplished it with what was left in their mouths.[35]

"Pinching and Pricking" are common complaints in the Salem Papers. Depositions regarding at least twelve people speak of the strange pains. There were no actual pins; it was the painful sensation they complained about. (Pins were commonplace in Salem Village, but none would have been left lying about—in that era pins were two-piece handmade constructions and very valuable.) Sarah Bibber, thirty-six years old, reported that she had experienced terrible "beating and pinching and almost Choaking me to death and pricking me with pinnes after a most dreadfull maner," along with her pitiful four-year-old child who had been so tormented by pain that it was struck "into a great fit" and writhed violently so that her husband could not even hold the child.[36]

As with the other symptoms, not only females complained. William Brown, a seventy-year-old Salisbury resident, reported that thirty years past, in 1662, his wife Elizabeth, "being a very rasional woman & sober & on that feard God," had been bewitched. A witch had begun tormenting her with sensations such "as birds pecking her Legs or pricking her with the motion of their wings and then would rise up into her stomach with pricking pain as nails &

pins." This painful experience caused her to "bitterly complain and cry out Like a woman in travail [childbirth]." Finally she would get a painful lump in her throat "in a bunch like a pulletts egg" and scream at hallucinatory visions of witches choking her. Brown continued with the rest of his sad story: a few months after her affliction he had returned home to an irrational Elizabeth who would not "owne him, but said they were divorst." He was able to calm her, but from that time to "this very day" she had been under a "strange kind of distemper and frensy, uncapible of any rasional action." Of course, he had consulted Doctor Fuller and Doctor Crosby, but both said that her "distemper was supernatural & no siknes of body, but that some evil person had bewiched her."[37]

Brown was accompanied by a friend, Robert Pike, who swore that Elizabeth Brown had been "a Rasional woman before she was so handled," and that many knew of her present condition and could testify to the truth that "shee yet remaines a miserabl creetr," who had lost her mind.[38]

Fits, pinching and pricking, swollen throats—none of this would have made much difference to history if the afflicted persons had not also experienced visions. *They* were more puzzling, more frightening, and extremely difficult either to interpret or to understand. Many colonists reported seeing spectral images, such as unrecognizable people, small animals (cats, rats, dogs), and visions of bright lights. Other than fits, complaints identified as visions or hallucinations were the most frequent problem people experienced, according to the seventeenth-century witch trial records. Some suffered only a mild affliction, perhaps a heavy weight on their chests and legs while in bed, which left them momentarily paralyzed, or fleeting pains in their abdomens, arms, or legs. But the hallucinatory vision was usually part of every affliction. Naturally the power of suggestion and the group dynamics of behavior would lead us to believe that several accounts may have been exaggerated or simply invented. Examining the bulk of the complaints, however, one appreciates the sense of seriousness about these experiences. Victims had nothing to gain from relating their experiences in public; they were not suing for damages. As to gaining stature, money, prestige, or affection—no one would receive anything of value in exchange for a made-up or exaggerated story, so fabricating evidence against others was in no way lucrative. In fact most people were not quite sure whom or what they saw, and had difficulty interpreting their visions. In spite of embarrassment and a certain element of public disapproval (as well as the chance that those speaking up in any way might themselves be labeled witches), people participated by revealing their experiences because they felt so vehemently about bringing the problems to an end. If a legal solution could be found, they were willing to tell their story. It was a matter of right versus wrong; witchcraft was both a crime and a sin. Useful citizens did their part by revealing their own torments and speculating about who might have caused them.

Young girls, while not the only victims, were the ones most often brought to court or taken from house to house for examinations by doctors, clergy, and other afflicted individuals. They were unable to prevent their exhibition for the dozens of onlookers and "witnesses" who viewed their fits and seizures. Many men reported similar circumstances, but they were not dragged into public view, nor were they physically examined by probing hands. Allowed a more respected position in society, they did come forward and talk about previous afflictions, particularly those that had happened years ago. None appeared while in the throes of convulsions except for John Indian.

James Carr told how he had been "taken after in a strange manner" twenty years earlier (1672 or so), "as if every living creature did run about every part of my body ready to tear me to peaces and so I continued for about 3 quarters of a year." He had consulted Doctor Crosby, who had treated him with a "great deal of physick," but that did not help. He tried a stronger medication, tobacco tea (probably made from lobelia), but it too failed. Then he told James that he believed the man was "behagged"—a term meaning bewitched—and asked him who he suspected wished to hurt him. James had been reluctant to name names, but the doctor pressed him, and the two discussed their opinions of several local women as potential causes. Nights later James had more hallucinations, this time of a cat which he attempted to strike at but could not because he was partially paralyzed. A short time later the doctor's "physick" began to work on him, and he was cured.[39]

William Stacy of Salem Town, thirty-six years old during the trials, related how he had been sick with smallpox fourteen years earlier (about 1678), at the age of twenty-two, and after that he had had night visions of a person in a black cape and hat, as well as a bright light "as if it had been day." He reported falling down several times, unable to walk correctly. Sadly he related how his daughter Priscilla, a "Thriveing Child," had died in 1690 after she "suddenly Screaked out" and continued in a series of fits for two weeks before she "dyed in that lamentable manner."[40]

Samuel Gray, a middle-aged Salem resident, told how he had experienced hallucinations fourteen years past, in 1678, while in bed. He had awakened and had seen the house "light as if a candle or candles were lighted in it," and he saw a vision of a woman wearing black, whom he believed was a witch. Worse, she had done something to his child, who had been "thriveing" but afterward "did pine away and was never well, althow it Lived some months after, yet in a sad Condition and so dyed." When the night vision returned to him on a different night, he was able to recognize it as a neighboring woman, Bridget Bishop, who was subsequently convicted and hanged for witchcraft.[41]

Stephen Bittford, a young man aged "about 23 years," testified that in April 1692 he too had been strangely afflicted. He had been staying in a neighbor's house at Salem when he began having "very grate paine in my neck and could not stir my head nor spake a word," which continued for two or three days, during which time he could not "stir" his neck, even though he could move the rest of his body. He had visions of what he thought were two local women (who were already accused of witchcraft regarding others), but he admitted that he could not "say that it was they that hurt me."[42]

John Hughes was another male who admitted he had had a strange experience that winter. He had seen a vision of a huge white dog, then later while in bed "in a closed room and the door being fast" so that nothing could possibly come in or out of the room, he had awakened to see a "Great light appear in the said Chamber." Rising up in his bed, he saw a "large Grey Cat" at the foot of the bed. While not as physically exhausting or remarkable as the violent experiences of the young children, Hughes too knew that things had not been normal.[43]

When Mary Warren was tried for witchcraft, she was in a particularly difficult position: she had been one of the afflicted persons herself and was now on trial as a cause of oth-

ers' afflictions. The difficulty of defending herself upon in-
terrogation must have been extremely problematic, because
she frequently lapsed into catatonic, speechless states, fell
to the floor in seizures, or was otherwise disabled. She did,
however, talk openly about the apparitions she had seen
when she had been "Afflicted" weeks earlier: "I thought I
saw the Apparission of A hundred persons," rather than a
single witch. Because her "Head was Distempered" at the
time, she had no control over what she had said regarding
who was afflicting her. She admitted that when she was
"well Againe" she could not identify any of the apparitions
she had seen in her hallucinatory state.[44]

One of the last cases brought to court in Salem took
place in October 1692: Sarah Cole, a resident of Lynn, was
accused of bewitching her own niece and nephew. The two
children experienced pins "thrust into" them, nosebleeds,
"bites and scratches," and other fitful behaviors. So did their
mother, who was "sorely afflicted" and "saw a ball of fire,"
which disappeared. She also saw a vision of a dog which she
tried to strike with a spade, but the "Dog went out at a crack
in the side of the house."[45]

A noted vision occurred during church services, when
Abigail Williams and Ann Putnam, Jr., cried out, one after
another, that they saw a woman they recognized as Deliverance Hobbs, sitting up near the ceiling on a beam. When defending herself in court, Deliverance Hobbs had a difficult
time because she too had earlier been afflicted and could
not explain any of it. Having experienced the oddly unnat-
ural "torments," she did not disparage what the others were
going through; during her questioning at the bar, several afflicted people in the courtroom fell into fits. Her questioning went on and on, and she eventually revealed her vision
of a tall black man.[46]

Many others had what became well known as a "Black Man" vision: a strange man dressed in black—sometimes tall, often small—who meant to do them harm. It was a painful, terrifying vision, and one that horrified those who saw it. But there was another stylized vision common to many of the deponents. It was the vision of a large ball of light, or a person (not always a man) in white garments, sometimes a virtual choir of angels—a vision of a strange but beautiful place from which some people said they had not wanted to return. Was it the same sort of vision described by those who have been close to death? We will never know, but for those New Englanders who experienced the "White Spirit" vision, it too was a strange experience, not to be forgotten.

Bad enough that the colonists' health was suffering, but this occurred at a crucial time. The colonies were on the verge of entering a market-based, capital-driven economy, and they were losing their assets, for livestock too were suffering and perishing.

The animals' afflictions and symptoms have seldom been considered by historians, unless attempting to ridicule some hapless farmer. The depositions and court records are rife with documented livestock losses. Witnesses recounted how their precious animals had acted strangely, gone seemingly out of their minds, and often died. Before the era of veterinary medicine, animal owners must have tried folk remedies and practices handed down to them from generations of English herders and stock raisers. Combing through the court records, one can see how busy the colonists were, men and women alike, tending animals and bartering with each other for animals or their products. Whenever someone

visited a neighbor there was usually a reference to something they were either coming to buy or sell; animals and
their products were the mainstay of the economy, and the
loss of a family's assets was potentially devastating. There
were no banks to borrow from for replacements; nor was
there a large central marketplace where one could obtain
new stock. Families cherished and respected their "milch"
cow as well as their hogs, chickens, and goats. If animals
sickened and died, families were financially shattered.

The colonists reported symptoms in their animals that
were eerily like those of the "bewitched" people: bruised
flesh similar to the "bites" on the girls' arms, animals jumping and running in circles like the hypermanic children, and
animals with their necks twisted out of place, or tongues
hanging stiffly from their mouths. A biological pathogen was
affecting both people and livestock, but it was nothing they
had encountered before.

John and Mary Edwards described how their pigs and
yearling cattle had been afflicted in January, the very time
the Parris household experienced its first problems. The
Edwards' nine pigs were "taken suddenly," five of them
dying. Two weeks later three yearlings who had seemed to
be "very harty" and in good shape were also "tacken su-
dently" with unusual fits: "Jumping & Roreing till they
tumbled downe with in a Littel tim after one another." John
swore to the testimony relating to the death of the pigs and
cattle, and his complaints were taken as seriously as if they
involved a sick child or paralyzed maidservant.[47]

Neighbor after neighbor came forward to describe livestock problems: John Bly bought a sow from Bridget Bishop
that began foaming at the mouth and went crazy, running
and jumping for several hours. Goody Henderson told him
that the cows at Eastward had gotten the same sickness, and

the people there had cured them with red ochre and milk. Benjamin Abbott had not only been afflicted in his body but testified that "strange & unusuall things has happened to his Cattle"; some had died "suddenly & strangely" which he could not tell "Any naturall reason for." Others had aborted calves, and some of the cattle came out of the woods with their tongues hanging out of their mouths in a "strange & affrighting manner." He and his wife could give no reason for this except perhaps that it "should be the effects of Martha Carrier's threatenings." Martha Carrier happened to live in an unfortunate neighborhood, for her own nephew, a twenty-two-year-old wounded veteran of the French and Indian War, also testified about his cattle problems which he figured were Martha's doing. A three-year-old heifer, two yearlings, and a cow had all died mysteriously. The nephew could figure no known reason nor "any naturall Causes of the death" of the animals, so he felt it must have been his Aunt Martha's malice (her husband had gotten into a fist-fight with him in the past—not a close family relationship by any means).[48]

Another of Martha's neighbors lost a cow "in a strange manner"; it died on its back, legs in the air. It was spring, and the animal had appeared "very lusty"; then another cow that was "well kept with English Hay" and in excellent condition "pined & quickley lay downe as if she was asleep & dyed."[49]

Mary Eaton charged Sarah Cole with bewitching her cow, which had been "taken in a Strange maner," and Abraham Wellman accused Sarah of causing his gentle cow to be "taken with fitts," and impossible to milk. John Cole, Sarah's husband, reported so many hallucinations in the night that he admitted he had been unable even to sleep in his own house for several nights, as he was always "sorely molested" by visions of a great cat, a ball of fire, and other unusual apparitions.[50] He too suspected it was all his wife's doings.

The largest losses were reported at the trial of Sarah Good by Samuel Abbey, a middle-aged man of Salem Village. About 1689 or so, in winter, his cattle began dying, and he lost several in an "unusuall Manner, in drupeing condition." They were often unable to rise off the ground and stand on their own. Their lackluster behavior grew worse until the Abbey farm had lost seventeen head of cattle and some sheep and hogs within two years. At the same trial the Gadges testified that their cow had perished in the same odd way, in a "Sudden, terible & Strange, unusual maner," and that neighbors they had consulted thought it must be witchcraft. After all, they had "opened the Cow yet they Could find no naturall Cause of s'd Cowes Death." These were people who lived side by side with animals their entire life and who examined the animals, both exterior while alive and interior postmortem. (Of the bewitched animals that were examined after they died, there is no mention whether or not the meat was eaten.)

The strange animal deaths, particularly cattle, extend over time, from the 1660s to the 1690s, just as the human complaints do. Back in 1669, while living at Amesbury, John Kimball's cattle herd had been struck in the spring in a similar fashion. He found one lying in the pasture stark dead with her head to one side; a little while after that another cow died and then an ox, and then several other cattle. Kimball had been helpless to cure the animals or determine what was wrong with them, as they had appeared to be "stout, Lusty" cows. John Houp, of nearby Ipswich, told of cattle in the town that were "Bewitched to Death, leaping three or four foot high, turning about, Squeaking, Falling, and Dying, at once." Several others also testified about live-stock afflictions and losses. In Amesbury, Susannah Martin was accused of bewitching cattle because the dead animals exhibited no sign of any livestock illness that colonists rec-ognized. One man testified that the deaths of his cows one spring had amounted to a loss in assets to the value of thirty pounds. Not only health but wealth was being lost, and doc-umented in court.[51]

At Elizabeth How's witch trial, her own brother-in-law testified that he believed she had been the cause of his ex-tensive cattle losses over the past several years. One of her neighbors, Isaac Cummings, related how his horse had died a mysterious sudden death, with bruises on the soft inner hooves and the appearance of being burned on the face. Timothy Perley and his wife claimed that Elizabeth had not only been responsible for the deaths of their cattle but had caused their daughter to be "destroy'd by Witchcrafts." The girl had tried to fling herself into the fire and the river dur-ing fits.[52]

At Martha Carrier's trial, Sarah Abbott and her husband testified that the accused witch had made Mr. Abbott "Af-flicted in his Body," and his cattle had died at the same time from unaccountable causes "such as they could guess at no Natural Reason for." Alan Toothaker, a veteran of the Indian Wars, John Rogers, and Samuel Preston all came forward at Martha Carrier's trial to accuse her of their cattle losses. The animals had been "Thriving and well-kept" cattle, who had succumbed in "strange Deaths, whereof no Natural Causes could be given." Cotton Mather called Martha "the Ram-pant Hag" and accused her of causing the by now familiar symptoms: pain in hands and chest, paralysis in half of the face and legs, "burning" pain in the stomach, loss of hearing, mental "distraction," and, in many cases, death or lifetime "insensibility."[53]

Inevitably, someone from nearly every household was involved that spring and summer as Salem and the surrounding area tried to make sense of what was happening to the residents and their animals. Contrary to popular conceptions of the Salem events, not only young girls were involved, and the craze was not due to mere jealousies or attention-getting behaviors. And the accusers were not disgruntled housemaids or jealous wives. One of the hearings involved the death of Daniel Wilkins of Salem Village, who had been "taken speechless" in mid-May of 1692 and could not speak for sixteen days. Twelve men attested to how he had suffered and how they had tried to help him. He had been unable to eat, so they had summoned "the French doctor," but his response was that "it was not a naturall Cause but absolutely witchcraft to his Judgment." They carried two of the other afflicted victims up to Wilkins's house to see if they could make out what was wrong with him, but he died three hours later in a "most dolful and solomne Condition." They had examined Wilkins's body, at the behest of the town constable, John Putnam, and had discovered several bruises on the back, and the skin broken and "many places of the greatest part of his back seemed to be prickt with an instrument about the bigness of a small awl." One side of his neck and ear were badly bruised too. When they turned the corpse over, blood had streamed out of the nose and mouth. The body did not swell and did not "purge else where," which was like nothing they could understand. No diarrhea, no vomiting, just hemorrhaging through the mouth and bruises, which meant bleeding under the skin.[54]

There was no way to make sense of things, but hard-

working, community-minded, God-fearing English men and
women testified in court that, "whereby great hurt and damage hath been done to the bodys of said persons," the community therefore *"Craved Justice."*[55] If it was not a medical
problem, then it had to be witchcraft, which made it a legal
problem. The victims of crime demanded justice. In the
seventeenth century, justice may have meant something
else entirely.

CHAPTER THREE

The Response

> I thought that I had greatly neglected my duty to
> my Children, in not admonishing and instructing
> of them; and that God was hereby calling my sins
> to mind, to slay my Children.
> —John Goodwin, 1688[1]

Initially the response by the families of those afflicted was
to seek medical help. The records mention many consulta-
tions with physicians and trials of various healing tech-
niques: smoking tobacco, brewing tobacco tea, bloodletting,
bedrest—anything else that might work. Animals were examined internally if they died strangely, but usually nothing,
except bruised flesh, could be found. Doctors of that era did
not routinely perform autopsies. Family members might
agree to it or request it, or community members might agree
to it in suspicious cases. Physicians would not have
"opened" anyone who was of stature in the community; the
impact would have been too distressing. They could, however, perform autopsies on bondservants or slaves.

Giles Corey, one of the few men actually tried for witchcraft (several other men were accused but not prosecuted),
was brought forward to discuss events that had occurred seventeen years earlier, when he had "kept a man in his House,

that was almost a Natural Fool." The mentally ill man was perhaps a bondsman, or Giles may have taken him in, providing room and board in exchange for work. The unfortunate man had died, and the physician (whose father happened to be an esteemed physician as well as the first governor of the Bay Colony) had performed an autopsy. The man had been "bruised to Death, and having clodders of Blood about his Heart"; the death looked extremely suspicious and was still well remembered in the colony. Corey had not been charged with murdering the man because there were no real signs that he *had* been murdered. But no one had been able to figure out the cause of his death.[2]

Deodat Lawson had been pastor at the Salem Village Church from 1684 to 1688, before the arrival of Reverend Parris. He knew everyone in the area and recognized the significance of events, which he wrote about in his book, *A Brief and True Narrative*, finished in April 1692 when the trials were just beginning.[3] He named the "afflicted" in attendance on March 21, 1692, at the public examination of Sarah Good for witchcraft. There were four married women (Pope, Ann Putnam, Sarah Vibber, and an "Ancient Woman" named Goodall); three "maids" (Mary Walcott, Elizabeth Hubbard, and Mercy Lewis); and three girls (Betty Parris, Abigail Williams [Samuel Parris's niece], and young Ann Putnam), all between nine and twelve years old.

This was a fairly large sample for a small village. The community of Salem Village, where events took place, was separated by a few miles from Salem Town, the original settlement. About six hundred people inhabited Salem Village, living in clusters of houses surrounded by at least an acre or two for gardens, pastures, and orchards. The number of people complaining of mental and physical problems was inordinately high in a community of that size. The public initially responded by attending preliminary hearings, to determine what direction to take, then ten days later a public fast was held on account of "these Afflicted Persons."[4]

Fasts were uncomplicated solutions to complicated problems of public health. The region had endured a history of epidemical sickness that people were not familiar with (such as the influenza-like disease that had killed so many New England children in the summer of 1675), attacks on houses and farms by French and Indian raiders, and periodic crop failures. War and famine they could understand. Disease— that was something the Lord himself had a hand in, afflicting the residents when they were not behaving properly or for some unrecognized spiritual failing. Fast days were ordered when the community needed to pull together over an issue no one could rightly explain. It was a community effort; a fast day was a "Day of Publick Humiliation, with Fasting and Prayer . . . that we may set ourselves sincerely to seek the Lord rending our Hearts and not our Garments before him . . . that the Lord may turn from his fierce Anger, that we perish not." Everyone was ordered to "forbear servile Labour" on that day in order to "apply themselves" to proper observation.[5]

Fasts were frequently held whenever there was an outbreak of disease, and in that regard they were fairly effective. People remained in their homes, the markets were closed, and workers were given the day off. This had a significant effect on controlling the spread of communicable disease, and for that reason was continued. In Salem Village, by the end of March, there were evidently so many afflicted individuals that suspicions must have arisen about something highly communicable in the community. People were taking no chances. A fast was ordered.

Contagious disease was well known in the colonies, as it

was in Europe at that time. The first arrivals at Plymouth had been delighted to discover that the Indian population had already been wiped out by an epidemic (now thought to have been introduced by the French in Nova Scotia), which they viewed as a providential act of God in their behalf. The pattern of life of the English colonists was usually a fairly good insulation against epidemical disease. Because of geography and social structure—they lived on scattered farms and in small villages—and because the population was fairly stable, intrusions by disease-carriers were lowered. People lived their entire lives in the same house or village; population density was low, and most people had little contact with the outside world. In villages and towns along the coast, where incoming ships brought disease-ridden goods and passengers, ships were quarantined. But that could not remove the problem of pirates unloading offshore onto isolated beaches, or rat-infested goods washing ashore to be retrieved by residents. Warfare brought disease, too. Captives taken by the French and Indians were held in jails and hospitals in Quebec and later returned. Those who survived had often developed an immunity to a particular virus and shed the virus, infecting their communities, when they returned.[6]

Some diseases were always present, others were introduced at intervals when the population's immunity had weakened or disappeared. Year after year colonists experienced "winter" and "summer" sicknesses, which were affected by climate and other factors, such as the presence of standing water for mosquitoes, or domestic livestock for ticks.

Smallpox in particular was devastating because it appeared at intervals, between which a large number of non-immune people were born who seldom escaped when an epidemic returned. Inoculation (called variolation, or even "buying the disease") was discovered very early in China, introduced to the Muslim world, and by 1720 practiced in England and her colonies. In the seventeenth century, however, it was not yet a reality in New England. Disease was thought to be God's punishment for sin, and if one was to die from it that was the will of God. Artificially introducing the disease into a cut on the skin was considered a heathenish practice, no doubt reinforced by the knowledge that African slaves knew about and participated in it.[7]

A major smallpox epidemic hit Boston in 1666, perhaps imported from England by ship, though it had existed among the French and Indians in Canada for at least the preceding five years, and Indians and traders may have brought it down from Canada to New England colonists. It spread that summer and increased by winter. After receding into a "trench" phase, as many communicable diseases do, it disappeared, then reappeared just over twenty years later. In 1689 and 1690 it was widespread in Canada and New England as far south as New York. It spread quickly through Boston during the winter of 1689 and was thought to have come from a ship of infected passengers from the West Indies. Colonial troops developed smallpox while attacking Quebec in 1690 and brought it back to New England, too. Smallpox was by that time recognizable and well known by its skin lesions. The first public fast was ordered for Boston on March 6 to try to stop the spread of the disease. On July 10 another fast was ordered. The epidemic was widespread and fatal. Salem was affected too, and the town records of 1690 show that the community's leaders provided some food and care for the sick.[8]

Yellow fever was also familiar to colonists. Boston harbor had quarantined ships for months in 1648 and 1649 to keep the disease "at bay" due to a series of epidemics that had ravaged Yucatan, Barbados, and Cuba. No one realized that mosquitoes carried the sickness, but quarantine and public fasts seemed to prevent the spread. There were other disease epidemics too, but those who recorded the severity of epidemics relied on the number of fatalities to gauge the illness's threat. If the sick recovered, no matter how widespread the infection, it was usually disregarded. Epidemics like the one a traveler to Boston witnessed in 1671, where people were afflicted with "griping of the guts, & fever & Ague, & bloody Flux," were usually disregarded unless they produced significant mortalities.[9] New Englanders had experience with disease, but what they saw as "afflictions" evidently did not resemble smallpox, ague, or any of the other illnesses they had experience with.

The major disease in the colonies was malaria, the leading cause of sickness and death. One did not develop immunity to malaria but rather became reinfected over and over, weakening over time. It is debatable whether malaria was indigenous to North America or imported from Europe or Africa. During the eighteenth century, its virulence increased in the rest of the colonies while it steadily diminished in New England, where by the time of the Revolution it had practically disappeared north of New York City.[10]

In any case, English colonists had plenty of historical experience with epidemics and plagues. By 1692 much had been written about the London plague of earlier times. Medical writers, such as Richard Mead in his 1744 *Discourse on the Plague*, written eighty years after the last London plague, were sure that plague was carried by "goods"— items that were bundled or wrapped and then shipped from infected places. Like a perfume that holds its scent, goods retained their contagion and were brought into the body through inhalation. To counteract the problem, goods had to be unpacked and exposed to fresh air for forty days before being unloaded in port. The disease, which was believed to float in the air above the goods, could be monitored from time to time by allowing small birds to fly over the goods to see if they were yet safe to handle. Birds were thought to be highly susceptible to the plague because there had been such heavy wild bird mortality during the plague in England—"the Country has been forsaken by the *Birds*; and those kept in Houses have many of them died."[11] Among contaminated goods, cotton shipments were particularly suspect, as they came to England from Egypt and Turkey, countries highly suspect for plague transmission; eventually England would get much of its cotton from a pristine environment: the southern colonies of North America.

Since disease was thought to rise up in the air, bonfires had been ignited in the streets of London in an effort to combat the plague by taking the bad air up with the smoke as it rose. Fire also warmed the damp air that was thought to be partly to blame. Londoners ordered fires in the streets for three days, but four thousand people perished in one night before they realized their "mistake." They were *heating* the air with the flames, and the overheated air was recognized as the real problem. So the "thick air" was attacked by continuously firing guns into the air, but that too failed when people realized that the air was only becoming more overheated and that the black smoke and powder were doing nothing to make the air fresh and cool.[12]

So, looking back, people in 1692 could chuckle at the wrongful ideas of the earlier generation. Medical practitioners adopted the idea that the air was not as important as what was inside the body: it was imperative to keep the blood from "inflaming." While cool blood might not protect a person from disease, it made any illness easier. How did one achieve cool blood? Eating acid fruits: pomegranates, oranges, lemons, sour apples, and, for the person of limited finances, drinking vinegar. Interestingly, these are all sources of vital nutrients that would have helped alleviate scurvy and perhaps other inflammatory diseases.

Another technique to regulate the blood was the practice of phlebotomy or bloodletting. Mead wrote that many physicians began the treatment on the first day of a person's "Distemper" by bleeding about twelve ounces, then taking away four or five ounces more every two hours afterward. Plenty of cooling liquids and "Drinks" were given to the patient in between bloodletting sessions. Mead admitted that the English did not use the practice as aggressively as did French doctors, and admonished that perhaps the English should "draw Blood with a more liberal Hand . . . if we expect Success from it." After all, he pointed out, ". . . in so desperate a Case as this, it is more prudent to run some hazard of exceeding, than to let the Patient perish for want of due Evacuation." Lancet in hand, the English colonist was prepared to protect his health and that of his family by engendering a chronic case of anemia. Perhaps that is not all bad, because researchers in our time have found that low levels of iron-rich red blood cells create a less than optimum setting for viruses and bacteria to flourish. When epidemics like the plague and the 1918 flu struck, the "healthy" people had more severe infections than the anemic. Nutritionally deficient blood simply does not lend itself to a parasitic relationship with some diseases.[13]

Before 1750, medical care in the English colonies was provided by four different kinds of practitioners:

1. Political leaders: governors, administrative officers, and clergy. They rendered medical aid in their communities because they were the only ones with any substantial education.

2. Physicians formally trained in England. At that time they did no actual dissecting and cutting—barbers were relegated to "letting blood" (which meant surgery and amputations), as it was forbidden by the church.

3. Local men who had apprenticed with older practitioners and took over their practice. They "rode" with a doctor while doing their apprenticeship. Bondservants and slaves were allowed to practice medicine after an apprenticeship.

4. "Quacks and charlatans." They practiced "kitchen physick" or copied Indian healers.[14] The term was a way to denigrate women and men who attempted to practice herbalism on their own, without the seal of approval of the church or the local physician. Fraud was also a problem in an era when credentials, licensing, and even establishing identity were nonexistent. Desperate people will always turn to those who promise health. It was the same in the colonies.

What about midwives? They were present in Salem and in colonial settlements, but there are no references to individuals or their families seeking them out for assistance in combating an epidemic. At least there is no mention in the court records or diaries. Their domain was obstetrics, and the problems in Salem appeared to have no connection to that practice. Some women who were tried as witches had performed as midwives, but they were not tried for any offense connected with their vocation.

Cotton Mather addressed the issue of physical illness, or what he called "pure Distemper," but he felt that the people were experiencing something too odd and unusual to be connected with any known disease. He was a man of medical training, but he considered that perhaps the "Evil Angels" had taken advantage of the "Malignant Vapours and Humours" of the diseased body, perhaps preying on females more than males for the same reason the "Old Serpent" had accosted Eve in the Garden, rather than Adam.[15]

People's hallucinations and frightening visions were what caused Mather and the medical doctors the most consternation—while they knew how to treat skin blisters with oil, and how to wield a lancet, what about the continuing complaints of the afflicted that they were being roasted on a spit, or burned with irons? Since the cause of these symptoms appeared to be invisible, there was no way to counteract them.

Mather was as well educated as anyone about medical matters in America, and he pointed out that he had read "not a few of the best systems of Physick that have been yet seen in these American Regions." But nothing helped him figure out the "Name of the Natural Distemper whereto these odd symptoms belong."[16]

Whether it was the Black Man hallucination or the White Spirit hallucination, Mather and his cohorts had no means of explaining why people were experiencing these unusual visions or how they were to be interpreted. Mather had read widely about European witchcraft (multitudes of chapbooks about the witch-hunts had been published for a voracious public in Britain and Europe), and he noted that the bewitched in Sweden had also related seeing a "White Spirit" that comforted them in their miseries. Mercy Short and Margaret Rule, young women he had observed at close hand, had both experienced this vision; Margaret described the spirit's "bright, Shining and Glorious Garments." Mercy Lewis had seen the White Spirit vision too, describing it as a "man in white, with whom she went into a Glorious Place."

In her fits she said there had been "no Light of the Sun, much less Candles, yet was full of Light and Brightness." She did not want to leave the glorious place to return to reality. Several other afflicted individuals also related the White Spirit vision.[17]

While Mather had documented this vision, what could he do with it? How did it fit into witchcraft beliefs that focused on the evil spirits? That the people were seeing a positive, uplifting vision was beyond his ability to grasp. He explained to the crowd at a Salem witch execution in August 1692 that the five individuals condemned to the gallows deserved to be hanged. George Burroughs, one of the condemned, was a former minister in the very village. He moved the crowd with an impassioned speech before the noose was slipped around his neck. Mather, seated on horseback in order to better address the crowd, reassured the hesitant citizens of Salem Village that the pleasant visions many of the afflicted had experienced had really been the Devil who "has often been transformed into an Angel of Light." The people renewed their vengeance and resumed the executions. Burroughs and the other convicted witches—John Proctor, John Willard, George Jacobs, Sr., and Martha Carrier—were executed and their bodies crammed into rock crevices—arms and legs sticking out uncovered.[18] It was a horrible sight that Salem would never forget.

Did other regions experience the strange afflictions common in Salem? If so, how did they interpret what they experienced? Connecticut, a nearby colony with similar geography, terrain, and settlement patterns as well as cultural roots, provides a good example for comparison. Events had begun in Connecticut as far back as 1668, when Thomas Tracy, thirty-one, had been in bed with his brothers (sharing beds in colonial homes was commonplace; travelers were even expected to share with strangers while staying at inns) when he began experiencing strange hallucinations that he was being strangled and the "flesh being pulled from his bones." The rest of the family tried to help him, but he could not speak; in the morning light they were able to see the marks of the pinching. That same year Mary Hale had awakened at night with a terrible weight on her legs, and then "it came upon her stomach and oppressed her so as if it would have pressed the breath" out of her body. She had hallucinations of a big dog in the room, and suffered pain for days that "made her fingers black and blue."[19]

Domestic livestock was afflicted in Connecticut, just as in Salem; one man lost four calves and thirty lambs in two weeks—they had "seemed to be well and were dead before ye next morning." Henry Grey had a calf that "roared very strangely" for six or seven hours. It had scours and was lame in a "very strange maner," the calf "being well and ded [dead] in about an hour." No stranger to domestic livestock health problems, Grey had examined the carcass, "and when it was skinned it lookt as if it had been bruised or pinched on ye shoulders."[20]

Afflictions and unexplained physical and mental symptoms continued in Connecticut sporadically until 1692, when Salem's problems peaked. That same year twenty-four-year-old John Barlow also experienced a terrifying nighttime paralysis and speechlessness, pinched feet, and saw a specter while the dark room was "light as day." Sarah Bates reported being called to Daniel Westcott's home to cure his French maidservant who had been "taken with strange fits." She found the girl in bed, and thinking it was an illness from "sum naturall cause," Bates advised the

Westcotts to burn feathers under her nose—an accepted curative that had "dun good in fainting fits." The next day
Daniel Westcott came for Bates again; when she arrived she
found the girl in bed "seemingly senceless & spechless, her
eyes half shet," but her pulse was regular. The mistress de-
cided to have her bled. Bates was reluctant to do it "in her
condition," but the mistress insisted. They bled the girl
from her foot, after which the girl cried out and broke into
hysterical laughter. Over the next few days her violent fits
increased. She claimed she was being pinched by unseen
hands, so a visitor examined her with a lantern and saw a red
spot on her neck "as big as a piece of eight," which turned
"blue & blacker than any other part of her skin." Two more
marks appeared on her shoulder before she fell into another
fit.[21]

Goodwife Clausen was accused of bewitching the maid,
and at her trial Daniel Westcott related how the girl had become afflicted: "Being in ye fields gathering of herbs, she
was seizd with a pinching & pricking at her breast"; she
went home and fell to the floor crying "with her hands
clasped," which continued for two days. Then she began
hallucinating about a talking cat. The fits increased and cats
began trying to kill her; her torments continued for thirteen
days, peaking with her experiencing "about 40th" fits—one
during which she did somersaults "two or three times heels
over head" as well as appeared to be breaking her own neck
with a stiffened arm. Instead of being taken seriously, as
those in Salem were, the relatively few afflicted individuals
in Connecticut, such as the Westcott's French maid, were labeled liars and frauds, and ignored. Goody Clausen was
found not guilty.[22]

Interestingly, the Westcotts' eldest child, Johannah, had
suffered fits at the same time. She developed paranoia, hys-

teria, and hallucinations—"some neighbors advised a removal of her," so the parents sent her to stay with a family in Fairfield, after which she was reportedly "getting over it."[23]

Connecticut records indicate a similar spate of incidents affecting people and livestock as occurred at Salem Village, but there were fewer incidents and they were not as terrifying to the community. In 1662, thirty years before the outbreak at Salem, the last execution for the crime of witchcraft in Connecticut had been held. When afflictions occurred in 1692, people went to court, but in some cases those who accused another of bewitchment were charged with slander and brought to court as well. Due to the fear of slander suits (and expensive judgments), there may have been many more cases of afflictions in Connecticut that were not brought to court or identified in the record. Like the Westcotts' daughter Johannah, the afflicted individuals may simply have been put in an upstairs chamber or sent elsewhere to live for a time.

In Salem, things were dealt with head-on. Aggressive in combating any efforts to hinder the building of a shining city on a hill, community members would not let problems rest. At the same time there were so many more afflicted individuals in Salem that the problem could not be ignored. Members of the medical community had resigned from taking action against the afflictions because their efforts had failed to relieve the sufferers. They deemed the symptoms to be so unusual as to be supernatural, essentially passing the problem on to religious leaders. From that point the problem was viewed as a spiritual one, and the ministers became involved.

The two most influential religious figures in the community were Cotton Mather, minister of the South Church of Boston, and Samuel Parris, minister of the small church lo-

cated in the center of the controversy, Salem Village.[24] It was
in Parris's own household that the first recorded afflictions
in 1692 erupted in Salem Village. Mather had also studied
the problem at close hand, taking on the Goodwin children's
case years earlier. Interestingly, both men turned first to
medical authorities and were reluctant to believe that ailments and afflictions within their own congregation, household, or family could be attributed to witchcraft.

   Samuel Parris is an interesting individual, viewed as either a pitiless madman or a McCarthyite in Puritan's clothing, depending upon the historian. He responded to the
witch-hunts through the pulpit; the sermon was the primary
medium of communication of the time. Community affairs
were handled from the pulpit through sermons, which became the media of the day. During a crisis, the minister's
sermon was highly important to the community and how it
dealt with the situation. The minister's role was to quell fear
by interpreting events rather than dictating policy. He interpreted events for the community by sifting them through a
sieve, in this case, of religion. One of Parris's sermon note-
books has survived, allowing us a glimpse into what Salem
residents heard from the pulpit during this crucial time.

   The second son of an English merchant, Parris had en-
rolled at Harvard but dropped out in 1673 when his father
died and left the elder brother the family business and hold-
ings, leaving Samuel a plantation in Barbados. Arriving in
the West Indies, he discovered that his estate was only
about twenty acres, and the region was rife with a violent
slave rebellion. Black slaves had plotted to overthrow the
English and kill all the "Masters, Mistresses & Overseers."
The plot had been discovered and several slaves executed,
and "a bloody Tragedy" averted. In August of the same year

a violent hurricane hit Barbados, destroying buildings, killing many people, and beating ships to pieces in the harbors. There were "many lying in their Beds of Sickness" who were hurt by the violent storms. Sickness, in the form of malaria, was another problem for the English in Barbados—at least for those without the genetic condition known as sickle-cell anemia, which protected many African slaves taken to the island against malaria. The English had no such protective adaptation and hence suffered miserably, dying like flies whenever an epidemic of the sickness passed through. Barbados was in such miserable straits—"Such another Blow will bring Barbadoes near the Horizon"—that the Parrises did not stay long.[25]

Parris moved his family to Boston, where he tried other ventures, finally turning to the ministry, the career his older brother had pursued before he inherited their father's large estate. He was freshly ordained in 1689 when he arrived at Salem. His niggling bargaining with the townspeople regarding his salary, the amount of firewood he should receive, and whether or not he would get title to his house have been criticized by historians, yet his actions were justified. In many New England communities the economy was slack, and the churches (and the minister's family and household) had to be supported by the community. There were no schools to speak of, nor a local tax structure, so community members were left to hammer out working agreements with their only public servants, the ministers. For Parris, this negotiating was a fact of life; he had been a merchant and businessman and knew he had to bargain—detail was important to him, and he had no doubt seen what happened to businessmen when they ignored the details and terms of a contract. His sermons, too, showed attention to this characteristically New England sort of attitude; he was, after all, in the land of entrepreneurs, talking to them in their own language.

Parris was not an experienced minister when the witch-hunts hit Salem. He had only a few years of village experience to rely on, and he was overwhelmed by the responsibility that a full-fledged witchcraft outbreak placed on a minister's shoulders. He was unethical (he used other people's sermons) and in way over his head. Moreover, his own household was hit by the afflictions; his daughter Elizabeth, niece Abigail Williams, and servant John Indian all fell into fits. Williams and Indian were eventually tried as witches in spite of their afflictions. Parris had little impact on stopping the witchcraze or even protecting his own household.[26]

His notes for almost every sermon repeatedly use the phrases "wounds & bruises." He cajoled the congregation into realizing that the colonists' sufferings were similar to Christ's, that "afflictions indeed are common to all" but particularly belonged to true believers. In 1690 he had preached that sufferings were not the Devil's handiwork but were punishments handed out to believers by a loving God. He put the question to them: "How may we know yet our bruises, wounds & afflictions are from divine Love?" He answered for them: by the soul's unity in God and man, "in & after afflictions."[27]

Parris's words allude to a growing problem in Salem Village: so many of the residents were sick—many of them congregation members who were true believers—that it became difficult to maintain the credibility that all punishments were coming from God. In one sermon Parris aimed at the many members of the congregation who had been accepting communion while unworthy and not in a state of grace. He told them that because they were not truly worthy, their behavior (and sin) had brought them "losses, pains, sicknesses, Death & the like." He chastized them that because they insisted on coming to the Lord's Table in a state of sinfulness, "for this cause many are weak & sickly among you, & many sleep." Their unworthiness and sin was the reason "they are weak feeble & sickly & Sad in Soul, as well as in body."[28]

Sickness as a result of sin was a popular theme for Parris that spring. In April he reiterated that because of their sin, many in the community were "Weak & Sickly & many sleep." He continued, "for this cause the Lord smites you with sundry plagues viz: with weakness, sickness, yea, & death itself. Plagues & punishments are the worthy fruits of unworthy communicating at the Lord's Table." Parris proved to be an ineffective leader at a time when leadership was everything. Perhaps the situation in his own home caused him consternation, which added to his own inexperienced and lackluster leadership. While the trials raged, he sent his daughter away to help her find a cure elsewhere. His niece and slaves, John and Tituba Indian, were afflicted and accused of witchcraft. Meanwhile his own wife pined away, sick and bedridden during the entire span of events. She died after a lengthy illness in 1696 at the age of forty-eight. If Parris had been a more forceful voice in the community—had been more experienced in leadership—he might have taken events in a different direction.[29]

The other spiritual guide and interpreter was twenty-nine-year-old Cotton Mather, the minister and prolific writer. Before novels, before even newspapers appeared in the colonies, Mather cranked out books, pamphlets, and letters, circulated and sold them, and exported them to London. He was an avid reader who leaped at the chance to write about witchcraft, having read many of the books and chapbooks turned out after the witch-hunts of Europe had grabbed readers' attention. He knew he could market books about colonial witchcraft (after all, his father, Increase, had done well with his book *Remarkable Providences* in 1684; the sections about witches had perhaps been the most popular with readers). Cotton was the right person at the right time to research firsthand and write about events as they unfolded. He studied the Goodwin children of Boston and wrote about them, then was brought in as a consultant on several other cases when children were afflicted in Boston as well as Salem. His book *The Wonders of the Invisible World* was completed, printed, and sent to a London publisher by October 1692, the year of the trials. Essentially about the Goodwin children and their afflictions, it was an account of the events at Salem with the goal of proving witchcraft and the "Devil discovered." His next book, *A Brand Pluck'd Out of the Burning,* was based on afflictions experienced by Mercy Short from the summer of 1692 until her "deliverance" in March 1693. Mather considered Mercy's case to be identical to the other "Bewitched peple" in the Boston-Salem area.[30]

Mercy Short had been taken captive by "our Cruel and Bloody Indians in the East," who had killed her parents, a brother, and a sister, and taken three brothers and two sisters along with Mercy into Canada. She had been returned from Quebec to Boston with other prisoners who were "redeemed"—ransomed from the French—and returned by ship to Boston on November 19, 1690.[31]

She was seventeen years old and employed as a servant in Boston when she was sent on an errand to the Boston jail. Several prisoners happened to be incarcerated there on witchcraft charges, so that later, when she too became bewitched—exhibiting similar "Fits as those which held the

Bewitched people then Tormented by Invisible Furies in the County of Essex"[32]—people deduced that she had been bewitched during her visit at the jail. She experienced myriad afflictions over several weeks, including being unable to eat for twelve days. She recovered, but by the winter of 1692–1693 she began to exhibit fits again along with an inability to eat for nine days. She lay stiff in bed for weeks, her eyes wide open, "moving to and fro," but she could not see anything. When Mather and other watchers tried to elicit an instinctive reaction by feigning to hit her in the eyes, they could not make her blink. They tried putting their hands over her open eyes to keep her from seeing the "invisible Fiends" that troubled her. At the same time they found that she could not hear—they tested by shouting "extremely loud in her ears," but she did not respond.

She hallucinated about a Black Man with eyes like bright balls of light; she felt pinches and pinpricks on her skin (Mather saw the "bloody marks" appear); her throat was so sore she claimed to be swallowing pins, and her jaw was rigidly set open for hours at a time.

While Mercy endured this trauma, Mather and six others continually experimented on her—slapping, holding their hands over her eyes or mouth, screaming, lifting and propping her, continually trying to figure out how the Devil was working. Obviously they had no fear of contracting whatever affected her. She screamed and begged for salad oil to ease her throat pain—they responded, and the oil seemed to help, causing her throat swelling to subside.[33]

Mather observed how long Mercy went without eating (up to fifteen days at a time) and compared it to the writings of Henricus Ab Heer, private physician to royalty in France, who had noted early in the seventeenth century the manner in which a bewitched girl, brought to him for a cure, had gone fifteen days without eating. Mather had read Robert Plot's *Natural History of Oxfordshire*, which described a young bewitched woman in England in 1671 who existed "without Eating or Drinking for Ten Weeks altogether." Mather concluded that "Long Fasting is not only Tolerable, but strangely Agreeable to such as have something more than Ordinary to do with the Invisible World." Mercy's afflictions reinforced Mather's beliefs that New England was being attacked by devilish torments already recognized in England and continental Europe. Mercy also exhibited blisters on the surface of her skin as well as hallucinations of being roasted on a spit and of having a hot iron thrust down her throat—her tongue and lips had the skin "fetch'd off."[34]

Eventually Mercy "underwent another sort of plague," which Mather had seen in only one or two other bewitched persons: she became hyperactive and frenzied, and her "frolicks" were those of a "wild-cat." She failed to recognize the spectators and became "witty . . . Insolent and Abusive" to them. They tried to circumvent the devil by praying aloud, reading Scriptures and singing hymns loudly, with "Half a Hundred in the room." They even resorted to slashing the air in the room with swords to try to strike the specters that were hurting Mercy as she lay bedridden.[35]

Mather found it remarkable (and perhaps fortunate for his research) that six months after Mercy Short's seizures subsided, another young Bostonian, Margaret Rule, began to exhibit the same afflictions. (He noted that Rule had not known Short, nor what her behaviors had consisted of.) On September 10, 1693, Margaret Rule fell into odd fits at Assembly and had to be carried home and confined to bed for six weeks. Soon the observers (thirty to forty people) saw black and blue marks on her skin, places where pins had pricked her, and saw her joints distorted into "astonishing" convulsions. She did not eat for nine days, largely due to the fact that her muscles had frozen and her jaw was clenched shut.

Perhaps because he was young and good-looking (and the subjects were usually young girls), or perhaps because he was taking on such a controversial subject as witchcraft, Mather was impelled to defend his writing as being of historical value to the church and the world—and it has been. His first-person chronicles of what the afflicted people experienced help us view events in a detailed way. Mather could easily write today; he was by turns a health writer, horror writer, and tabloid reporter, all with dollops of religion for acceptability. He published more than four hundred pamphlets or books marked by "piety, learning, pedantry, vanity, and great defects of judgment," according to the historian J. Franklin Jameson.[36] In many ways that accurately describes his writing: but his defects of judgment were overshadowed by his grasp of the market for books and the way he kept his finger on the pulse of colonial culture. Not only did he select topics in which the general reading public was intensely interested (or at least interested enough to buy books), he sprinkled his writing with words uncharacteristic of a minister: "Surprising Things," "Remarkable Encounters," "An Account of Wonderful & Surprising Things," "Remarkable Salvations," and "Remarkable Disasters." All were used purposefully to gain attention, pique curiosity, and stimulate interest. Mather was really a *writer*, and not nearly as interested in being a minister.

In his book *Decennium Luctuosum*, or *Sorrowful Decade*, he wrote of a "Surprising Thing laid before the Reader for him to judge, (if he can) what to make of it." He described the witch era as the "story of the Prodigious War, made by the Spirits of the Invisible World upon the People of New-

England, in the year, 1692, hath Entertain'd a great part of the English World, with a just Astonishment." He continued, "I have met with some Strange Things, not here to be mentioned, which have made me often think, that this inexplicable War might have some of its Original among the Indians, whose chief Sagamores are well known unto some of our Captives, to have been horrid Sorcerers, and hellish Conjurers and such as Conversed with Demons."[37]

Mather's efforts to write history as it was happening have left us with excellent detailed narratives of the events of the day as they affected individuals. But he too could not understand what was happening within his very family. Although he had seen people afflicted by unknown disease or spirits, when his own wife's health was impaired he did not link it to what others had experienced. In autumn 1700, eight years after the Salem events, she suffered from "sore throat and such Tumour and such Dolour and such Danger of Choaking, and such Exhausting of her Strength with it, as is not common," that Mather knew she was in trouble. Two months later she gave birth to a son who was so sickly at birth that Mather knew the boy would not live long. The infant, named Samuel, had "more than an hundred very terrible Fitts" before dying at the age of two months. Mrs. Mather remained bedridden for months; many possible remedies were tried, even some she saw in hallucinatory visions, in which a "grave Person" (someone from a cemetery or the afterworld) told her to use sheep's wool to stop the pain in her chest. Mather described his wife of sixteen years as a "Gentlewoman of a melancholy Temper," and her death was hard on him. During her illness, three of the children and a maid came down with smallpox. They recovered, but the maid was so "distracted" she was useless; Mather had to dismiss her.[38]

CHAPTER FOUR

Mental Illness and the Persecution of Witches

/\

From that Time, this poor [Mrs.] Whetford was utterly spoilt, and grew a Tempted, Forward, Crazed sort of a Woman; a vexation to her self, and all about her; and many ways unreasonable. In this Distraction she lay, til those women were Apprehended by the Authority; then she began to mend and upon their Execution was presently and perfectly Recovered, from the ten years madness that had been upon her.
—Cotton Mather, "Wonders of the Invisible World"[1]

Two different topics seem to meld at this point: witchcraft and mental illness. And one is compelled to examine the historiography of each. Both come together in the Salem story as well as in many other historical events.

Studying the history of medical psychology, or psychiatry, requires a different approach from studying the history of medicine and surgery. Although surgical advances may be studied by looking at the stories of individual doctors and their endeavors and innovations, the study of the mind requires an investigation into the entire culture, including its legal system, religion, and philosophy—branches of learning that had dealt with matters of the mind for centuries—not simply the field of medicine, which dealt more with humors of the body, blood, and herbal antidotes.

Mental illness has been around a long time, even though misunderstood and largely unrecognized. Hippocrates wrote about the "sacred disease," which is thought to have been a form of mental disorder, perhaps epilepsy, and which the Greek medical profession considered to be caused not by the hand of God but by some form of physical illness. Other disorders, characterized as melancholia or mania, appear in medical writing from early times. Plato wrote on such topics, dividing mania into two distinct types: one caused by divine origin, the other due to physical overexertion. In classical times, mental disorders were not clearly the domain of the medical community because thinking was affected; therefore psychoses and disordered thinking belonged to the realm of the philosophers.

In the third century Marcellus, a physician, wrote about lycanthropy, a disease that caused people to wander in deserted places, such as cemeteries, at night where they howled. By the fourth century, mental disorders were more prevalent and varied. Besides the occasional werewolf, society had to deal with many more psychoses and unusual behaviors. The ideas of the classical thinkers were abandoned as society turned toward militarism during the age of feudalism. Science, knowledge, and medicine took refuge in monasteries, where they were subject to theosophical interpretation.

Between the fourth and seventh centuries, attitudes turned against science. Mental disorders were determined to be the work of the devil. Treatment became the application of sainted relics and incantations as exorcism. People tried to help the "possessed," to rid them of demons.

Epidemics of mental illness across western Europe began with the "children's crusades" of 1212, when thousands of French children followed a shepherd boy after a vision of Jesus appeared to him and gave him a letter to take to the French king. The numbers swelled as the group marched—eventually more than fifty thousand children and youths planned to take back the Holy Land—but they were sold into slavery by the Moors.[2]

John Howells, in *The World History of Psychiatry*, suggests the possibility that "in the Middle Ages episodes of collective psychopathology took place in Italy on a small scale, similar to the great epidemics of St. Vitus's dance, of flagellantism (obsessive whipping of oneself), of lycanthropy (the belief in shape-shifting into wolf form), often under the influence of peculiar religious sects and of superstitious beliefs." The Flagellants, members of a movement that appeared in the thirteenth century and spread within twenty years to Bohemia, Moravia, Poland, Italy, and even Russia as late as the seventeenth century, gathered by the thousands to wander from city to city, whipping themselves to rid their bodies of demons.[3]

Some disorders of behavior were not recognized as diseases at the time, such as the "dancing epidemics" in the Middle Ages, when people took to the streets in a state of hysterical agitation and many succumbed to epileptical fits. Such mass psychoses took place in Holland in 1373, where they were called St. John's disease. The Middle Ages also saw epidemics of "tarantism," a collective psychopathology which was blamed on the bite of the tarantula and characterized by acute depression and withdrawal. On a smaller scale, similar outbreaks occurred in monasteries and orphanages which were both refuges for the ill and institutions where numbers of people were brought into close proximity, engendering the spread of disease. Mass psychotic movements during the thirteenth and fourteenth centuries terrified the church because they could not be controlled.[4]

About the thirteenth century in Europe, recognition of mental illnesses increased and interpretation became more sophisticated. Being "troubled with visions" or being "fiend-sick" and "wit sick" started to appear in the literature. Miraculous cures were written up too, usually taking place at the tombs of the saints. Religious ecstasy along with a belief in miracles became part of the faith and culture of medieval life. When only a few people acted in an unusual manner, they could be assumed to be special in some way, but as the numbers of afflicted and recognition of the spread of mental illness grew, society changed in its attitude toward the suffering. Despite a long tradition of caring and comforting those who were sick, people met the larger numbers of those needing care with less pity and more fear. Sick people were no longer carried to visit saints' tombs or touched with relics or bones for a miracle cure. Insane individuals were turned out by their families to wander because people feared the devil more than they pitied the afflicted. Economics may have also played a part in these decisions, as communities could not or would not care for those who were unable to work. The ill wandered the streets, slept in stables, and lost their humanness. Their appearance and personality became animal-like, which led to increasingly prevalent superstition and religious persecution.

For Christianity, mental illness posed a dilemma. It was difficult to incorporate the mentally ill into the faith because many of those with problems were highly religious, though visibly abnormal. Eventually the church realized that all disorders could not be caused by the devil, because so many of the martyrs and saints had behaved similarly. Was the insane Christian a saint or a demon? In many cases the line was finely drawn. In the fourth century a woman, along with two other women, was "somnambulistic and suffered from attacks of *grand hysterie*" while in ecstasy during Mass. She saw revelations, conversed with angels, and claimed to have seen the Lord. She became a prophetess to bishops.[5]

The idea that physical illnesses were natural and that mental illnesses were supernatural gradually became more entrenched in society. Any sort of cure was elusive and based more on supernatural elements than antidotes. Medicinal practices revolved around using plants such as the peony, which was employed as a remedy against demons, sorcerers, and epilepsy; gold and silver were used to cure melancholia. As mental illness proliferated, simple herbal or mineral cures could not pacify communities reeling with seemingly insurmountable numbers of incurable individuals.

If illnesses could be divided into these two categories, there had to be a way to recognize and determine whether a disease was physical or supernatural. One method was to shout a biblical passage into a convulsing patient's ear; if it elicited a response, this proved that the situation was caused by demons. The holy words frightened the demon, allowing the patient to respond and proving supernatural possession. If, on the other hand, the person remained in a stupor during the shouted verses, the illness was considered natural and could be treated with herbals, minerals, or by bloodletting.[6]

In the fourteenth century, people sought to connect human illness with the natural world through astrology and the study of the first open cadavers in centuries. Some medical writers in the 1500s insisted that mental diseases were not supernatural but due to unknown organic causes. They spoke out against the mistreatment of the afflicted, yet they had no popular or clerical support, nor the understanding to recognize mental disorders in their entirety. It was an era when spontaneous generation of life was an accepted theory, considered proven by the appearance of worms in rotting flesh. John Lange, a medical writer in 1554, wrote about an autopsy on a suicide victim whose stomach contained a piece of wood, four knives, two pieces of iron, and a bunch of hair. He noted another woman he had seen vomit up two iron nails, two needles, and a bunch of hair. Rather than seeing the connections between the two as mentally related, he believed the objects had been put into their bodies by a diabolical trick; their presence in the women's stomachs merely confirmed to him the existence of supernatural disease.[7]

Although the scientific understanding of mental disorders remained minimal, the legal system found it necessary to define parameters of witchcraft, possession, and insanity. Courts justified their examinations of potential witches through the long-held beliefs put forth in a book, *Malleus Maleficarum*—known as the "Witches' Hammer"—a 1489 guide written by German Dominican fathers and approved for publication by Maximilian I of the Holy Roman Empire. It became the authority for witch identification. A hint book, it gave practical tips on how to recognize and treat witches, and it identified different forms of witchcraft manifestations and diabolical possession. So popular it went through ten editions in the next 150 years, it widely influenced people because it identified behaviors that could be brought to court for prosecution.*

*It is interesting to study the American Psychological Association's

The manner in which some mental disorders recede and reappear, leaving an individual coherent between bouts of insanity, was difficult to understand. Lord Hale, an Englishman, rose to prominence in the mid-1600s because he found a way to resolve this longtime dilemma. He had formulated a way to discern temporary insanity which the court could use during witchcraft or sorcery trials. Based on the accepted scientific ideas of his day, he claimed that the "moon had a great influence on all diseases of the brain," particularly dementia, which caused susceptible individuals to be "at the height of their distemper" in the "full and change of the moon, especially about the equinoxes and summer solstice." Such people, however, had sane intervals between the full and change of the moon, when they regained their sanity or at least competent use of their reason. This idea, based on Lord Hale's legal reasoning (rather than science), became the criminal law regarding temporary insanity. Lord Hale promoted total insanity as the only condition that excused criminals from prosecution. He himself, using this enlightened reasoning, had sent two "sane" witches to their deaths with a clear conscience.[8]

Witches were an intrinsic part of Europe's belief and folk system. Much has been written about the era of the witch-hunters and the grotesque tortures and executions that riddled the area, giving rise to the image we hold of the "Dark Ages," a time when fear and horror must have been rampant. But what are the facts? Robin Briggs has recently writ-

handbook for identifying mental disorders. Called the *DSM-IV* (*Diagnostic and Statistical Manual*), it comes in a pocket-sized version. One leafs through the lists of behaviors and tallies up a score in order to determine what mental disorder the patient suffers from. It has become today's version of the *Malleus*. Pierloot, *Health Medicine and Mortality in the Sixteenth Century*, 140.

ten an interesting account of the European witchcraft era.
By studying records from four hundred witch trials held in
Lorraine during the late sixteenth and early seventeenth
centuries, Briggs was able to piece together what really happened. Rather than the parallel often drawn today between
witch persecution and the Nazi Holocaust, it was never a
matter of genocide or of gendercide. Briggs notes that the
often-used explanation of the witch persecutions as efforts
by the ruling elites to manipulate society for their own interests is inaccurate. He also questions the hugely exaggerated
number of executions involved. A powerful mythology
claims that 9 million women were burned in Europe. But
the most reasonable modern research suggests 100,000 trials
between 1450 and 1750, with 40,000 to 50,000 executions, of
which 20 to 25 percent were men.[9] Particularly unpleasant is
that many of the outbreaks of persecution involved children,
as either complainants or accused witches.

In spite of the overwhelming belief in witches as a cause
of afflictions, some medical practitioners and researchers did
attempt to find and cure organic causes of mental disorder.
In the sixteenth century Johann Weyer, a revolutionary
thinker, insisted that "evil spirits" meant illness, and he promoted the groundbreaking idea of consulting a physician
when a person was in a stupor or possessed with a "spirit,"
even though he showed no signs of physical illness. Weyer
attempted to prove that the illnesses attributed to witches
came from natural causes, and he challenged the way that
witches and sorcerers were punished and executed. His
writings were listed by the Catholic church in the next issue
of the Index Librorum Prohibitorum, and Catholic commu-
nicants were prohibited from reading it until the early twentieth century.[10]

In Saxony, an area heavily inundated with witch problems, the criminal code noted the appearance of several books which argued that sorcery was not a crime but a superstition and a "melancholy" (depressive mental illness). Observing that Weyer's book was among those to claim that witches and sorcerers should not be put to death, the authorities considered his writing "not very important, for he is a physician and not a jurist." Critics alleged he was probably a sorcerer himself, and his ideas only encouraged more sin, enhancing the "Empire of Satan." Weyer's influence at the time was negligible, but he kept his job as personal physician to Duke William of Julich, Berg, and Cleves. Unfortunately, when the duke had a stroke that left him mentally impaired, Weyer was suspected to be a sorcerer and the duke, now rather insane, his willing accomplice. In 1588 Weyer died, and his work was thereafter ignored.[11]

Other scientifically based opposition to witchcraft persecution was also largely ignored or suppressed. When Reginald Scot wrote *The Discoverie of Witchcraft* in 1584, he meant it to serve as a scientific unmasking. The book was ordered destroyed by King James I, who reasserted the tradition of demonology in his own publication, *Daemonology.* Superstition and religion as explanations for human frailties were so entrenched in the cultural system that medical professionals who attempted to prove otherwise risked their reputations and often their lives.[12]

Less than a century passed before physicians in Europe were regularly called upon to treat mental illness. By 1667 blood transfusions were used to treat mental disease. A French physician "let out ten ounces of blood from a vein in the arm and let in five or six ounces of blood obtained from a leg artery of a calf" in a thirty-four-year-old man who had the transfusion to cure mental disease. By the third day "the patient had quieted down and his mind had cleared," and he completely recovered. All the professors of the École de Chirurgie attested to the satisfactory outcome. The practice was immediately supported by physicians in England and Germany, particularly as a cure for melancholia.[13]

Since Europeans had been utilizing physicians to treat mental illness, why did the colonists insist on witchcraft as an explanation for their mysterious ailments? In the seventeenth century the ability of men and women to act as healers was difficult for the Puritan community to accept, and the church continued to remind people that illness was an act of God. This religious stance on science and medicine meant that much of what could have been the subject of medical practice was relegated to the status of witchcraft. Church and civil authority were threatened by both good and bad witches. Good witches, who were able to cure the ailing poor, were a direct threat to the established hierarchy of minister over congregation, lord over peasant, man over woman. This was the principal threat of the "witch" to the church. Because of this power struggle, religious authorities as well as government actually resented the *good* witch more than the bad. The power to cure bodies (and hence souls) was the domain of God . . . and his agents.[14] This may help explain why no one in Boston or Salem took notice of the White Spirit visions. Cotton Mather, in his voluminous writings, ignored them, as did others at the time—at least those whose writings survived, or who wrote at all. It was later, in a similar context, that another minister in a small town saw his parishioners' visions as positive messages sent from heaven—and quite possibly began the Great Awakening.

In Salem, George Burroughs, in his own defense against accusations of witchcraft, quoted a book by the foremost critic of witch-hunts in England at the time, Thomas Ady. Ady's book, *A Candle in the Dark* (1655), questioned where in the Bible it was stated that a witch was a murderer or had the power to afflict with disease or infirmity. Ady asked where in the Bible did witches have imps sucking off their bodies. Women accused of being witches were frequently inspected by matrons and midwives to see if there were any strange protuberances where the witch had suckled "familiars"—demonic animals. Efforts were made to question biblical authority for the persecution of witches, but the arguments failed largely because reason plays a small role in the beliefs that motivate mass movements. The ability of authority either to stimulate or to break mass movements of persecution is crucial in determining the outcome.[15] Nazi Germany serves as a contemporary example. In late-seventeenth-century Massachusetts, authority was weak and inconclusive; when popular opinion swung behind witch-hunts, there was no stopping them.

Thomas Szasz, writing in *The Manufacture of Madness*, points out that this "human tendency to embrace collective error—especially error that threatens harm and commands specific protective action—seems to be an integral part of man's social nature." So when people were faced with questions about either witches or mental illness, they tended to embrace and preserve the popular explanation, which helped to consolidate the group. Accurate observations of the issue, which tended to divide the group, were negated, ignored, ridiculed, or punished. People cling to the popular ideas of their day because they confirm their ideas about life and themselves. They tend to reject anyone who refutes what they believe to be true. Szasz reminds us that it is important to pay careful attention to the dominant worldview in particular historical periods—it explains how people regard and conform to their physical surroundings, their society, and one another.[16]

Although Enlightenment ideas were spreading, and witchcraft as an explanation for mental illness would soon be abandoned, isolated colonial towns in early America had not reached that level of sophistication. There were no schools in Salem, not even a newspaper in Boston at the time. But people of all cultures and ages create explanations for phenomena they do not understand. Because the idea of sickness due to sin had not entirely disappeared, people employed it, and the paradigm continued for centuries. The idea that sickness, including mental illness, was someone's fault continued to shape thinking in both religion and science; when a shift in thought appeared in the eighteenth century that placed the blame on the individual rather than on outside forces, witches disappeared from public discourse.

In the eighteenth century, "alienists," a new breed of healer and authority figure, emerged. Syphilis was widespread, and it created a huge increase in numbers of the insane. It was a century mired in political and economic interpretations because no real scientific advances in medicine could be applied to psychoses; all blame rested with the individual. The division of the medical practice between physicians, barber-surgeons, and apothecaries also prevented advances in clinical medicine.

In 1733 Doctor George Cheyne, a Fellow at the College of Physicians in Edinburgh and a member of the Royal Society, published *The English Malady; or, a treatise of nervous diseases of all kinds, as spleen, vapours, lowness of spirits, hypochondriacal, and hysterical distempers*. It was based on the author's own case and was a turning point in the recognition that mental illness could be accepted without shame. Cheyne explained his purpose in the book's preface: "What I pretend to have done in some Degree in the following

Treatise is That I hope I have explain'd the Nature and Causes of Nervous Distempers (which have hitherto been reckon'd Witchcraft, Enchantment, Sorcery, and Possession, and have been the constant Resource of Ignorance) from Principles easy, natural and intelligible, deduc'd from the best and soundest Natural Philosophy; and have by the plainest Reasoning, drawn from these Causes, and this Philosophy, a Method of Cure and a Course of Medicines specifically obviating these Causes, confirm'd by long Experience and repeated Observations, and conformable to the Practice of the ablest and best writers on these Diseases."[17] Cheyne's writing came a generation after the incidents in far-off Massachusetts Colony; perhaps he was influenced by what had occurred there.

Another form of treatment arose that was based on the ancient observation that serious illness can sometimes be alleviated by contracting a less dangerous illness. Hippocrates suggested that convulsions could be cured if the patient came down with a quartan (a fever that recurred in four-day intervals, usually malaria). Galen also wrote of mental illness cured by a quartan fever. Even Hermann Boerhaave, the leading physician of the time, noted in 1724 that mental illnesses that failed to respond to other treatments were sometimes cured after a bout of tertian or quartan agues. In the late seventeenth century, physicians observed a case of "stupidity" which was cured as a result of an illness accompanied by fever. They observed that severe physical illness sometimes made mental symptoms disappear. These observations led to what was called "fever therapy" or treatment with inoculation of malaria or induced convulsions. Belief in the curative value of malarial fever was so strong that King Louis XI of France begged to be afflicted with malaria so that he might be cured of his epilepsy.[18]

In 1887 medical writers urged that this "experiment of nature" be intentionally imitated and, because psychoses were sometimes healed by other infectious diseases, that malaria be used clinically. This was not practiced at the time (at least not sanctioned, though some practitioners did inject typhus into patients experimentally), but in 1917, coincident with the encephalitis lethargica epidemic, such treatment was tried. Julius Wagner von Jauregg, of Vienna, experimentally inoculated insane soldiers with malaria. In one study in which nine patients were inoculated with malaria, six of them definitely improved, and four years later three of those were still able to perform their regular work. This treatment stimulated interest in treating general paresis (slight or partial paralysis) because remissions seemed to occur and to last longer than with any other therapy. The success was so striking that von Jauregg was awarded the Nobel Prize in 1927.[19]

Beginning in the late nineteenth century, Sigmund Freud steered mental health studies in a new direction. Childhood trauma and dreams were revisited for nearly a century as people explored their own past, searching for others to blame for their troubles. Freud himself ignored the encephalitis lethargica epidemic that swirled around him, writing: "the neuroses of our own . . . modern days take on a hypochondriacal aspect and appear disguised as organic illnesses." Freud's teacher, Charcot, had believed that witchcraft was a problem of neurosis—another aspect of neuropathology. Freud postulated that the witchcraft phenomenon of the Middle Ages had been due to a splitting of the consciousness which resulted in hysteria. He admitted that the term "hysteria" was merely a semantic one, but he applied it to what had previously been called "witchcraft." Early in his career Freud had found cases of hysteria in men, but he was denounced and refused by the Viennese Society of Medicine, which admonished him because everyone knew that hysteria resided in the uterus, and men could not have it. Unfortunately Freud was already in his seventies when Constantin von Economo's book, *Encephalitis Lethargica*, was published in the 1930s. Freud had been ill for twenty years with cancer. Had he been at the beginning of his career, and not an established, world-famous celebrity of sorts, he might have had different insights into the mind and human behavior.[20]

CHAPTER FIVE

The Forgotten Epidemic

Any important disease whose causality is murky,
and for which treatment is ineffectual, tends to
be awash in significance.
                                        —Susan Sontag

The mysterious afflictions that beset so many residents of
New England from the mid-1600s until the eighteenth century can be understood by looking closely at another time
when an epidemic with similar symptoms struck a modern
population. In 1916 physicians in Europe began to report
puzzling symptoms in patients: convulsions, hyperactivity
interspersed with periods of catatonic stupor, and severe
eye-muscle disorders. The illness left people mentally disturbed, in comalike trances (sometimes for years), or dead.
Between 1918 and 1920 more than half a million people
were killed outright by this severe neurological disorder;
survivors developed a panoply of Parkinson's symptoms that
lasted up to forty years. Called the "sleeping sickness," encephalitis lethargica, epidemic encephalitis, or von
Economo's disease, in ten years it spread to five million people worldwide. The disease was mystifying, of unknown origin at the time—and by 1930 had disappeared. (Encephalitis
lethargica is not the same as the African Tripanosomiasis, or

sleeping sickness, an insect-borne parasitic disease which is spread by the tsetse fly.[1])

Dr. Oliver Sacks, who worked with many of the survivors years later, said that "manifestations of the sleeping-sickness were so varied that no two patients ever presented exactly the same picture, and was so strange as to call forth from physicians such diagnoses as epidemic delirium, epidemic schizophrenia, epidemic Parkinsonism, epidemic disseminated sclerosis, atypical rabies, atypical poliomyelitis, etc, etc. It seemed, at first, that a thousand new diseases had suddenly broken loose."[2]

Encephalitis is one of science's great enigmas. It is basically an inflammation of the brain, which may affect the brain stem and can physically alter the gray matter of the brain. When the invading organism (virus, bacteria, or spirochete) enters the bloodstream, it centers in the brain where it causes inflammation. White blood cells invade the brain tissue to fight off the infection, which causes brain tissue to swell and can destroy nerve cells and cause hemorrhaging and bleeding within the brain. Symptoms of encephalitis include an initial headache and fever, followed by confusion, hallucinations, paralysis on one side of the body, memory loss, difficulty in speaking, drowsiness, possible coma, epileptic seizures, loss of hearing, sensitivity to light (photophobia), stiff neck, and difficulty controlling the eye muscles. Additional symptoms may include pupils of different sizes, double vision, and personality changes.[3] Viral encephalitis is unusual because of the polymorphism of the symptoms; every epidemic consists of varying complexes of characteristics, and the effects may appear differently in individuals affected by the same epidemic. Every epidemic of viral encephalitis produces its own specific syndrome and degree of severity.[4]

Encephalitis is not very common in the United States today; each year approximately fifteen hundred people are affected by it. But without public health practices in place, it could expand to epidemic proportions. Epidemics of encephalitis are caused by a wide variety of viruses as well as bacterial agents (rickettsia, which cause Rocky Mountain spotted fever) and spirochetes (responsible for Lyme disease). Encephalitis occurs sporadically too, and has been associated with chicken pox, measles, rabies, HIV, and Creutzfeldt-Jacob disease ("mad cow" disease). Those most at risk for infection are people living in crowded, unsanitary conditions; the elderly, newborns, and infants; and those who have suffered another illness (usually influenza) that has lowered their level of immunity and put them in a weakened condition. Mild viral encephalitis is common and may go unnoticed; severe cases require hospitalization, but there is no cure. A small percentage of patients suffer brain damage which impairs mental or muscle functions. Most people, except the elderly and infants, recover in about three weeks.

Between epidemics of encephalitis (as with many other diseases) periods of calm exist when the disease goes into a "trough" and disappears for several years. These gaps may last for several decades. Not only may there be a time lapse between epidemics, there may also be quite a long period of time in an individual case between infection and the first appearance of symptoms. The virus may persist for a long time in a victim's central nervous system without causing any symptoms, only to be reactivated at a later time, even after a period of years.[5] An example is sclerosing panencephalitis, a brain inflammation that occasionally follows measles and usually occurs in children. In most cases of encephalitis, recurrence may result from new infection of the brain, when the virus enters the bloodstream again. Recurrent encephalitis may be the result of the person's immune system attacking the body in a misguided attempt to fend off what it thinks to be a reinfection by the original virus.

Historically there have been other epidemics of encephalitis lethargica, as from the days of ancient Greece when Hippocrates wrote, "I see men become mad and demented from no manifest cause, and at the same time doing many things out of place; and I have known many persons in sleep groaning and crying out, some in a state of suffocation, some jumping up and fleeing out of doors, and deprived of their reason until they awaken, and afterward becoming well and rational as before, although they be pale and weak; and this will happen not once but frequently."[6] Hans Zinsser notes in *Rats, Lice and History* that one of the cases Hippocrates described may have been encephalitis lethargica: a pregnant woman experienced back pain, fever, headache, pain in the neck and hand, and loss of speech. She went into delirium and suffered from paralysis of the hand and arm.[7]

In Europe in 1580, the Continent was swept with a fever accompanied by lethargy that left victims with parkinsonian symptoms. An epidemic in London, from 1673 to 1675, was described as "febris comatosa" and was characterized by hiccups; others occurred in Italy in 1597 and in Germany in the 1500s, from 1672 to 1675, and from 1824 to 1826. In 1695, Albrecht of Hildesheim wrote an elaborate account of a twenty-year-old woman who suffered disturbance of the eye muscles, parkinsonian symptoms, double vision, strabismus, and other symptoms of encephalitis following an attack of "brain fever." Other episodes were noted in Sweden from 1754 to 1757. Minor epidemics of "coma somnolentum" with parkinsonian features occurred in France and Germany during the latter half of the eighteenth century, with reports of hiccups, twitching of muscles, uncontrollable muscle movements of the arms and legs, and tics. Other epidemics were noted in Hungary in 1889 and in northern Italy from 1889 to 1891. In Italy an influenza epidemic in the late 1880s preceded the nona, a severe sleeping sickness which left its few survivors reeling with Parkinson's.[8] Physicians reported cases of encephalitis in Europe in the years 1903, 1907, 1908, 1910, 1912, and 1913, though none had become epidemic in scope. Rare cases continue to be reported today, but they have been largely sporadic. In 1976, fearing an epidemic of swine flu (which might have triggered or been followed by an epidemic of encephalitis), the U.S. government ordered the production of 200 million doses of vaccine, but fortunately the epidemic never materialized.[9]

The last pandemic of encephalitis lethargica swept the world from 1915 to 1930. The first reported cases were in Romania in 1915, France in 1916, and Vienna in 1917. The disease appeared in England in February 1918; within the next four months 230 cases were reported in London alone. By the winter of 1918–1919 it had spread to the United States, and in the following year it appeared in Canada. Within a few years it had spanned the globe, becoming pandemic in nearly every region of the world. Seven thousand patients were sickened in Japan, and thousands more in Australia, Russia, Uruguay, and India. In the United States the peak occurred between 1920 and 1929. Statistics on encephalitis lethargica cases in the United States are probably misleadingly low because the disease was reportable— meaning doctors were required to report it to the local public health department—in only twelve states. Reports made in 1922 totaled 1,333 cases, and 1,822 for 1921. Overall, the epidemic was associated with a 20 to 40 percent mortality rate for those who reached the acute phase of the disease. In North America an epidemic of influenza had arrived from central Europe in 1918, preceding the outbreak of encephalitis lethargica. The 1918 influenza virus has now been identified as the H1N1 influenza A virus, probably of swine-human origin. Initially people believed that the sleeping sickness and flu were the same disease or had a cause-and-effect relationship; now we know otherwise.[10]

At the time the most pressing problem was determining whether encephalitis was contagious and how it was being passed. Person-to-person transmission was rare, and there were few outbreaks in schools, institutions, and other public places. In France, between January and May 1920, more than four hundred cases were reported, but none could be traced to direct personal contact; in New York nine hundred cases were studied in which there were no secondary cases in the immediate families of the patients. Sometimes it appeared that mothers had infected their newborns. Usually not all members of a household came down with the disease, but in some cases several members of the same family or household were afflicted. During the entire epidemic, contagion was inconsistent, so victims were not quarantined. In spite of a few person-to-person transmissions, the disease was not considered to be a communicable threat.[11]

In the 1918 encephalitis epidemic, incidence of the disease increased during the colder months of the year. That increase coincided with influenza infection the same year, but of more than one thousand encephalitis lethargica cases studied in 1923, only four had a history of influenza during the preceding six months. Similar observations were made by others. From 1916 to 1930 there were several waves of encephalitis lethargica, which were really a series of epidemics. With each one, a slightly different assortment of symptoms prevailed. While sporadic cases of encephalitis lethargica had occurred in winter and spring, most epi-

demics began approximately at the beginning of winter, and
the greatest number of acute cases occurred in the first quarter of the year—from midwinter to the beginning of spring.[12]
The incubation period was highly variable: anything from
one day to two months was typical. Constantin von
Economo, professor of psychiatry and neurology at the University of Vienna, was the first to identify or "discover" the
disease as a distinct form of encephalitis. He named it encephalitis lethargica because of its victims' propensity for
sleepiness.

Encephalitis lethargica struck people of all ages but particularly those between ten and twenty-one. Young people
were most susceptible, with a *preponderance of women affected.*
Most of the afflicted whom von Economo studied were between the ages of fifteen and forty-five. The elderly were
relatively free from the disease. Frederick Tilney and Hubert Howe, physicians researching the epidemic in New
York, maintained that there was a greater proportion of cases
in young adults in the 1916–1927 epidemic. This chart
shows the age incidence of those afflicted in the New York
study.[13]

Under 1 year	3
1 to 10 years	29
10 to 20 years	29
20 to 30 years	33
30 to 40 years	22
40 to 50 years	21
50 to 60 years	17
60 to 70 years	7

Von Economo, with other physicians, had been mystified
when patients began filling hospital wards in 1916 with mys-

terious symptoms that pointed to meningitis—a marked lethargy, slight flulike symptoms, delirium, and what appeared to be epilepsy. He studied the disease in Vienna during World War I, when both the epidemic and the war were at their height. He was able to continue scientific work as the war swirled around him. Other neurologists in England, France, and the United States were also investigating the disease, and there was cooperation as well as competition among them. It does not appear that postwar politics and attitudes affected recognition of von Economo's findings. Translated into English in 1931, his work has been available to the English-speaking community for some time.

Perhaps overshadowing his work on the organic causes of these symptoms were the theories of a fellow Viennese: during the peak of the epidemical brain infection, Sigmund Freud was concentrating on the interpretation of dreams, hypnotherapy, and the revisiting of childhood trauma. His work, perhaps because it was more easily grasped, was widely disseminated while von Economo's was ignored because no one knew what to do with it. There were no antibiotics, no electron microscopes, no ways to identify viruses until the mid-twentieth century. Scientists had not recognized that viruses mutated, nor could they identify them. Von Economo was a man wrestling with a problem without the means to examine or interpret it.

Von Economo devoted his career to encephalitis lethargica, largely due to his mother's mention of a similar epidemic she remembered from the 1890s, when a sleeping sickness called nona had affected many residents of northern Italy and Austria. He had been stymied by the patients he was examining in 1916 until she mentioned the similarities to him; from then on his career was focused on the physiology of the brain and the relationship between organic disease and psychoses. By working directly with patients in Vienna, he was able to describe case studies of individuals in minute detail. These narratives were used to arrive at generalizations about the disease. Von Economo, as well as other researchers at the time, found similarities between their descriptions of the strange new disease. They began to recognize a worldwide pandemic of unknown origin. Naturally the new malady was compared to existing medical problems of the day—polio, rabies, syphilis, schizophrenia, and hysteria were all considered as possible causes of encephalitis lethargica.

Encephalitis lethargica appears to be unusual because its symptoms may assume so many different forms. Every epidemic consists of symptoms unique to it, and within a single epidemic, patients may have varying symptoms. Some may have one or two, others the full gamut. Von Economo, in studying this baffling disease, discovered that symptoms might change from one form to another in the same patient. For example, a patient might start out catatonic and shift to hypermanic, or vice versa.[14]

Von Economo described three major forms of the disease: somnolent-ophthalmoplegic, hyperkinetic, and amyostatic-akinetic. The somnolent-ophthalmoplegic type usually began with chills, headache, and vomiting. It appeared as a mild flu, and the patient became dazed and confused. Minor pains sometimes occurred in the limbs. Eventually somnolence took over and the patient could no longer stay awake, falling asleep even while chewing food at dinner. The sleepiness sometimes disappeared quickly, leading observers to believe the patients had simply overexerted themselves. In severe cases the patient remained in a trance far longer—up to thirty years in the cases later studied by Oliver Sacks in New York City. Patients might also have partial paralysis of the eye muscles, double vision, blurred vision, or a sensation von Economo called "dazzling"—seeing a bright light of sorts. In 1916 many people had experienced hiccups, temporary deafness, attacks of giddiness and silliness, vomiting, inability to walk in a straight direction, and paralysis of the extremities. Patients were unable to move an arm or a leg, or at times even to sit up. They huddled or slumped to one side, or simply lay still and appeared to be in a "pseudo-sleep." Their muscles became rigid, their face a tight expressionless mask. In this particular type of encephalitis lethargica, convulsive fits were seldom seen.[15]

The hyperkinetic form of the disease was distinct from the sleep-based form in that it was characterized by hyperactivity and anxious restlessness. During the 1920s this form spread so widely that it was thought to be the most frequent form of encephalitis lethargica. It began with a few days of discomfort, sore throat, severe headache, vomiting, pains in the back or neck, and muscle weakness. There were also fairly violent pains in the arms, legs, hip, abdomen, or chest (or, variously, in all locations). Next came restless motor activity with general mental unrest and ceaseless physical activity. Von Economo described the behavior: "The patient tosses about in bed, pushes the blankets back, pulls them up again, sits up, throws himself back again in a wild sort of haste, jumps out of bed, strikes out aimlessly, talks incoherently, clucks his tongue, and whistles—this unrest lasting for days and nights without a stop." Even if the patient seemed to be acting somewhat normally, behavior was never quite usual. One of von Economo's patients wanted a glass of water, lifted it halfway to his mouth, then poured it on the floor.[16]

The psychomotor activity lasted a few days or increased

in intensity until the patients were in a state of delirium; then they experienced hallucinations filled with moving objects. Hallucinations were optical or tactile—the patients thought they were seeing things or felt them touch or crawl upon their body. They were terrified and driven into panic—which increased their hyperactivity into a frenzy. In some patients the delirious state was replaced by a hypomaniacal excitement—a cheerful restlessness with blurred consciousness, singing, shouting, whistling, even harassment of the people around them. During sleep these patients seemed to be in ceaseless motion, their bodies jerking in all directions, rolling, shouting, whistling, and noisily fidgeting. This went on all night, then as morning came they fell asleep and slept during the day, reverting to the mania when darkness came again. This hypomaniacal state occurred over a few days or several weeks. Patients with mild cases might experience only slight restlessness and insomnia that lasted for days or weeks, then disappeared completely. Perhaps those cases went largely unrecognized.[17]

A panoply of other symptoms of the hyperkinetic form included:

—Disturbance of the eye muscles, or the clinical term "oculogyric crisis." The eyes looked up or down or sideways, and seemed to be frozen in place in either an extreme position or what might appear mildly as staring. The pupils of the eyes might vary in size and remain unchanged regardless of the amount of light in the room.[18]

—Body movement in an uncontrollable random chorea—flicking, jerking movements that shifted from one muscle group to another. Victims sometimes became so violently active that they had to be restrained to keep them from injuring themselves. Von Economo said they often "roll about continuously, jerk themselves up, throw themselves down and end up with numerous abrasions and scratches." Hiccups lasted several days. Jaw muscles would stiffen and contract, holding the mouth clamped shut or open for periods of time. The tongue muscles were sometimes affected, the patients biting their tongues severely or for long periods of time.[19]

—Intense pain in the abdomen, so sharp it was sometimes mistaken for appendicitis.[20]

—The patient might be able to take only halting steps or be partially paralyzed.[21]

—In some patients the skin surface was affected by twitches: short, quick, nonrhythmical movements that repeated in rapid sequence. Cold air caused them to increase in number and spread over the body. A sequence of quick, short, fluttering of contractions of muscle bundles, they are quite common in the hyperkinetic form of encephalitis lethargica.[22] The victims of the disease in Salem would likely have described this aspect of the illness as being pinched by someone—a witch, of course.

—Another notable symptom of the 1916–1930 epidemic was comparable to those instances in Salem when people watched as witches' "bites" appeared before their eyes. A peculiar disturbance of the skin surface appeared in the 1920s, referred to as an erythemata (a redness of the skin caused by dilation and congestion of the capillaries, and surface hemorrhage). Colonists likely exhibited the same condition, whereby witnesses watched in amazement as marks appeared on a neighbor's or child's body that looked eerily like red, ring-shaped bites.[23]

—Sensitivity to suggestion: if the victims' names were mentioned, they fell into a paroxysm of unrest or echoed the activity around them.[24] This reaction would explain the manner in which afflicted witnesses behaved during court sessions in Salem, when wild activity broke out among the afflicted from time to time, sometimes interrupting the proceedings with the noise and furor.

The third type of encephalitis lethargica that von Economo identified was labeled the amyostatic-akinetic form. It was characterized by extreme stillness; patients would lie on their backs with closed eyes but not sleep. They understood questions but took a very long time responding; they moved so slowly that they appeared catatonic. Patients remained for long periods of time in any position they were placed in. Their facial expressions never altered, their eyes seemed unblinking. Any action would take place over a long time span; for instance, chewing food would go on for hours without swallowing. The motion itself was almost imperceptible.[25] General symptoms of this third form of encephalitis lethargica included:

—excessive salivation;

—a peculiar shine on the face, termed "greasy face," caused by overactive oil glands in the skin;

—retention of urine for long periods of time;

—eye muscle disturbances similar to the other types of encephalitis;

—sleepwalking;

—paralysis of swallowing and speech.[26]

These three basic forms of the disease were seen separately in patients, but von Economo also noted that they would appear in the most varied combinations. The disease lacked a typical progression; there were no stages that patients went through while either worsening or recovering. Anything seemed to occur in any order. Each successive epidemic wave during the 1916–1930 period seemed to exhibit a few unique characteristics, such as trismus (lockjaw) during the 1920s, and at other times pathological yawning, chewing, clucking of lips and tongue, and, most common, hiccups.[27]

Other symptoms seen in patients who had any of the three forms included the very common swelling and pain of the parotid glands—the ones that swell with mumps. Some acute cases exhibited a jaundiced discoloration of the skin. Among those patients experiencing diminished movement or greater tension, there was sometimes an unusual burst of activity during which they overcame their disability; some patients who could walk only with assistance could suddenly dance quite well if they heard a particular type of music, or in a sudden state of excitement they broke from their catatonic state and ran after and jumped aboard a moving vehicle.[28]

After initial symptoms subsided, patients often developed later behaviors that repeated with a ticlike recurrence of movement, such as yawning, clucking and sucking, swallowing, or hissing. Fits of yelling or yawning transpired too, but they were usually influenced by suggestion, for instance by hypnosis or command. Speech was sometimes affected, and the patient would utter words over and over, making no sense. This compulsive repetition is known as palilalia. These repetitive ticlike behaviors were highly suggestible: they occurred at once if one talked about them to the patient. Simply talking about the problem could bring it on.[29]

Von Economo observed that in the course of the disease it was common for the patient to experience sensations of heat, burning, and violent pain in the extremities. Sometimes behaviors were grotesque: a patient would open his mouth wide, twist his trunk and arms backward in the most extreme fashion, stamp at the same time with his feet, and utter a snorting sound. Such attacks were excruciating to witness and often frightened the patient.[30]

Children were affected by the disease, of course, and von Economo recognized distinct disturbances that affected them in particular: altered personality, hypomania with insomnia at night (while sleeping during the day), outbreaks of anger that he called "blind orgies of destruction," talkativeness, impertinence, and anti-social behavior. He saw what he called outright misbehavior, too: plucking other people's clothing, making faces at them, lying, stealing, marking graffiti, vandalism, running away, arson, and other "dangerous acts" which endured over a period of years. In one group of Danish schoolchildren who recovered from encephalitis lethargica in 1924, "Changes in morals and character are the features which primarily stamp the picture," revealing a striking similarity between the children's behaviors. After returning to school they had poor concentration, short-term memory loss, faulty perception, deviant sexual behavior, a general lack of interest in schoolwork, and what von Economo called their "lack of mental perseverance," which made them "backward."[31]

But aren't these common behaviors for typically misbehaving children? Von Economo argued that the children were different because they recognized that their actions were wrong; they were sorry and regretted what they had done, but they could not arrest their behavior. The question arises: how does that differ from insanity? Von Economo pointed to the motor unrest, the patients' realization that their behavior was inappropriate and their regret over it. He found that the patients did not identify themselves with their behavior. They felt they were being made to do it, as if they were puppets on a string.[32]

Von Economo was not alone in studying the disease. In the United States, Frederick Tilney and Hubert Howe, physicians working in New York hospitals during the epidemic there, collected many similar findings. They were, however, reluctant to use von Economo's term encephalitis lethargica, arguing that the name they selected, epidemic encephalitis, was more appropriate—because it was the patient who was lethargic, not the encephalitis. No matter. They found the chief characteristics of the disease in the United States to be similar to what Viennese patients had exhibited, particularly the multiple symptoms. They identified nine types of encephalitis lethargica, expanding on the three von Economo had recognized. Their types were: lethargic, cataleptic, paralysis agitans, polioencephalitis, acute anterior poliomyelitic, posterior poliomyelitic, epilepto-maniacal, acute psychotic, and infantile.

The lethargic type was characterized by long periods of sleep that was so deep the person could not be awakened. The cataleptic type was characterized by symptoms similar to those of a sixteen-year-old girl who experienced severe pain in her finger that spread to her arm, where it lasted a short while, then disappeared, leaving her arm slightly paralyzed. At that point she suddenly began to act irrationally—becoming violent, requiring restraint and sedation. Her family believed she had "gone insane." She quieted into a deep sleep and slept without opening her eyes for eight weeks. Her limbs and body were held in rigid positions the entire time. Her jaw muscles held her mouth rigid; her facial muscles were stiff and gave her an unchanging expression. When alert, she demonstrated that she had been aware of what was being said around her while she was in the deep sleep. After an illness of forty-seven days, she fell into a "terrible coma" and died.[33]

The paralysis agitans type was exhibited by a thirty-eight-year-old secretary who noticed she was excessively nervous and unable to sleep at night. She was depressed and

doing poorly at work. She began running rather than walking, and had a hard time stopping. She had tremors in one
hand and one leg, and a rash on her body that looked like
hives. Her voice and facial expression were unchanging, mo-
notonous, and slowed. She began having "hysterical attacks"
and was sent to a sanitarium. She never experienced
lethargy or sleepiness; instead she was restless, insomnic,
and hyperactive.[34]

Those with the polioencephalitic type suffered from ex-
haustion, speech aphasia, double vision, facial palsy, strabismus, and paralysis of the oculomotor nerves and muscles.
This form was very mild, and the inflammation affected the
central nervous system only slightly. Usually total recovery
occurred.[35]

Acute anterior poliomyelitic encephalitis was exhibited
by one four-year-old child in New York City who had a
"dragging of the right foot, and weakness of the right leg,"
and fell into a deep sleep that lasted fourteen days. Six
weeks later he was able to hold his head up and gradually
began improving.[36]

The posterior poliomyelitic type was evidenced by the
case of the thirty-nine-year-old woman who got out of bed
one January morning in 1919 and was unable to stand up.
She fell down and was barely able to crawl back to bed,
where she fell asleep. For the next three weeks she slept
most of the time. She awoke for meals but was uninterested
in anything. Her physician thought her condition was hysteria. She had double vision and a painful tingling sensation
on the left side of her face. On the same area of her face a
rash of herpes-like eruptions appeared. She began to feel
tingling and burning on her right leg. She never experienced
paralysis.[37]

Epilepto-maniacal and the acute psychotic types were

both evidenced by the thirty-six-year-old man who had influenza which led to a "psychic disturbance," making him "unmanageable." He hallucinated and suffered "ill-defined, unorganized delusions." They escalated at night, when he became violent and destructive and required restraint. He experienced persecution delusions which exacerbated his "maniacal manifestations," until he had to be kept under the influence of narcotics. After five weeks he went into epileptic-like convulsions that lasted several hours, with increasing frequency and severity. He died in a coma after six weeks of illness. Another thirty-six-year-old male had similar symptoms, especially at night when he went into "an acute psychosis" in which he "seemed to hear voices and reply to them." He became noisy and uncontrollable in the hospital ward, eventually lapsed into a coma, and died.[38]

Finally, the infantile type was studied at Babies' Hospital in New York City, where infants were found to have the disease at a very early age. One six-week-old stopped eating and passed into a deep sleep for two weeks. He survived because he was fed from a tube, but the left side of his face remained paralyzed. Two other four-week-old infants experienced the same symptoms and outcome: deep sleep for more than two weeks, eventually returning to a mild alertness. Both had the left side of their faces paralyzed.[39]

Not all the New York cases exhibited lethargy or drowsiness. Not all included fever, either. Patients who did not fall into a sleep seemed to be affected in the opposite manner, becoming extremely restless and hyperactive. Symptoms seemed as disparate and the duration of the disease as unpredictable as in the patients in Vienna. Some patients in New York died after a few days, some after several months; many cases with the most severe symptoms cleared up rapidly. Little connection was made to the influenza epidemic, which was rampant at the same time, since half the patients in the New York study had not had the flu. General symptoms seen in New York hospitals included restlessness, delirium, hallucinations and delusions, lethargy, stupor, and muscular rigidity. Some people experienced a meningitis-like neck pain. Any of the symptoms might be exacerbated by fatigue, which extended the period of recovery.[40]

Through postmortem dissection von Economo was able to distinguish the disease from any other. He thought the visible effects resembled poliomyelitis, rabies, Borna disease of horses, typhus, or distemper. But by actually examining the brains of patients postmortem he saw the highly visible manner in which the gray matter had been altered by the infection. But absence of infection or inflammation of the spinal cord distinguished it from poliomyelitis.[41]

Von Economo observed that the "psychomotor excitement, the deliria, and the choreic unrest with skin and face twitches may last several days, sometimes from two to three weeks or even longer. . . . The disease may subside with rapid disappearance of its symptoms, and complete recovery occur." Or there was a "temporary apparent recovery for a few days or longer, and after that period of rest a renewed outburst of the state of excitement." At any stage, "sudden death may occur."[42] One never knew how patients would react the next day, or even the next hour. Or if they would survive. The disease was a harrowing and confusing experience for patient, family, and physician.

The psychotic aspects of encephalitis lethargica are its most challenging, and for those afflicted and their families, the most disturbing. Von Economo listed the frequent disturbances he saw in patients: delirium, hypomaniacal states, amentia (senseless madness), dreamlike syndromes, and depressive-apathetic or catatonic-stuporose states. The delirium began at night with dreamlike visual hallucinations while the patient was in bed. Over time the dreamlike delirium began to extend into the daytime. At first, patients could be talked to and they would respond; in later stages they simply lapsed into hallucinations. Patients with the hyperkinesis form of encephalitis experienced the "severe delirious states with very lively and terrifying hallucinations and excitement leading up to outbursts of violence." The hallucinations were absolutely terrifying: the patient was often "afraid that his legs or head are going to be cut off, sees hovering skeletons dance round him, can hear the clattering of their bones, they prick him, and drag him about, and etc."[43] The patients in Vienna hospitals were experiencing very similar paranoid delusions to those the colonists in seventeenth-century Massachusetts had endured more than two hundred years earlier.

In the 1920s the disease came on quickly, with some patients dying twelve hours after exhibiting the first symptoms. Early warning signs in such cases were often scant; the disease began with a sudden attack or seizure followed by paralysis or epileptic convulsions, and terminated in almost immediate death. Such cases were rare, though. Even in the 30 percent of cases that led to full recovery, the lengthy process took weeks or months. Relapses occurred at intermittent intervals, and nearly all patients had some small problem that remained after recovery. There seemed to be no relationship between the severity of the patient's symptoms and the result. Apparently mild cases were often fatal, while serious cases sometimes recovered. Recovery did not mean one was free of the disease, however, as new symptoms might appear several years afterward. The illness was able to persist for years in the central nervous system without causing any symptoms—people thought they were cured or did not even realize they had the disease. It seemed to be reactivated for no apparent reason after an indeterminate time span and restarted into virulent form. Seventy percent of those who recovered developed symptoms later. Sometimes these symptoms were fatal, but many patients survived for decades. The most frequent symptom was a parkinsonian syndrome that affected 60 percent of those who survived acute encephalitis lethargica. This syndrome appeared less than five years later in 50 percent of the cases and less than ten years later in 85 percent. Ultimately there was no reprieve from these late-onset parkinsonian symptoms. By the 1950s thousands of post-encephalitic patients who had become infected in the 1916–1930 epidemic continued to require constant care in hospitals around the world.[44]

In London the mortality rate was 20 to 40 percent. Von Economo found the following mortality-recovery rates in Austria: 40 percent mortality; 14 percent complete recovery; 26 percent recovery with remaining problems, but able to work; 20 percent left chronic invalids.[45]

Psychiatric disorders were the most frequent long-term result of encephalitis lethargica. Manifestations that commonly followed recovery from encephalitis included obsessive-compulsive behavior, personality changes, mood disorders, and anxiety. Between 50 and 100 percent of the survivors were estimated to have psychiatric disturbances. The most noticeable effects were recurrent mania, episodic depression, and bipolar mood disorder. Twenty-one of the 135 post-encephalitic patients at Manhattan State Mental Hospital in New York attempted suicide. The sort of manic behavior exhibited after the illness generally included euphoria, grandiosity, distractibility, and "sexual indiscretion."

Some physicians reported that their patients appeared to be driven to act out, showing no euphoria or humor.[46]

The psychiatrist Karl Menninger studied the mental aspects of the disease in one hundred cases of mental disease at the Boston Psychopathic Hospital. The range of mental disturbances was expansive, but he classed them in four groups: delirium; dementia praecox (now called schizophrenia); other psychoses; and unclassified. Those classified with dementia praecox, schizophrenics, were by far the largest group. The large number of mental patients filling sanitariums in the years following World War I was probably due to infection by encephalitis lethargica.[47] Cases of "shell shock" exhibited in returning troops may have been encephalitis lethargica, in its post-encephalitic form, in the form of catatonia.

How did physicians treat encephalitis lethargica? Von Economo reported the use of bed rest, diet, nursing care, and whatever relaxants were available, particularly warm baths for the most psychotic cases. But physically strong handlers who could keep the patient controlled were essential. Restraints were used often; electroshock therapy and lobotomies were thought to be useful. Before the era of antibiotics or even dependable tranquilizers, treating patients was no easy matter.

One experimental treatment that appeared to work involved injecting turpentine oil into the thigh with a hypodermic needle. The patient had to be given injections of morphine along with it in order to endure the pain. The injection swelled and filled with pus a few days later; it was then drained and used to inoculate the patient. The matter contained the antigens of the pathogenic virus, and the result was said to be "very satisfactory," if painful.[48]

In the late 1950s researchers discovered that the parkinsonian brain responded to higher levels of dopamine and could be somewhat normalized with a drug called L-dopa, or levodopa. In the late 1960s Dr. Oliver Sacks worked with patients from the 1920s epidemic, treating them with large doses of L-dopa, which sometimes elicited short-term remission of the postencephalitic parkinsonism they had suffered from for thirty years. He points out in his book, *Awakenings*, that large amounts of natural dopamine are found in beans—notably fava beans, and that beans may have been used as a folk remedy for parkinsonism for centuries.[49]

Psychiatric treatment came into its own following the epidemic. Early on, many psychiatric symptoms of encephalitis and their treatment were controversial. S. E. Jelliffe, a prominent American neurologist and psychiatrist, interpreted the oculogyric crises and compulsions of encephalitis as anxiety-relieving mechanisms. He believed it was the ego attempting to avoid a difficult or dangerous situation. He tried to analyze the major symptoms of postencephalitic parkinsonism, including the tremor and rigidity, in relation to psychological and sexual realms. He favored psychoanalysis as the most appropriate means of treatment. A devotee of Freud, he wrote to him for advice on how to interpret the encephalitic symptoms in psychoanalytic terms, but Freud did not answer. The father of psychoanalysis, living and writing in the same city as von Economo, during the same epidemic, never offered any interpretations regarding encephalitis.[50]

Jelliffe's interpretations influenced many in the medical world, because he owned and edited the *Journal of Nervous and Mental Disease* (the official journal of the American Neurological Association) during the period 1902–1944, and also edited *Medical News* and the *New York Medical Journal.* He was able to direct most of the work published in these journals along his pro-Freudian biases. Any researcher whose work failed to interpret encephalitis lethargica's psychotic manifestations according to Freudian psychoanalysis was rejected for publication. Not everyone agreed with Jellife; at one point the entire advisory board of the *Journal of Nervous and Mental Disease* resigned in protest over the pro-Freudian slant. Some neurologists and psychiatrists endorsed von Economo's concept that psychotic functions and disorders could result from organic causes, but a larger contingent continued to promote analysis as the means for interpreting and alleviating post-encephalitic syndromes. The "forgotten" epidemic has surfaced over the years in the debate between psychoanalysts and those who accept biological reasons as causes of behavioral disorders.[51]

What causes encephalitis lethargica? The 1916–1930 pandemic is still not well understood. The cause was uncertain at the time, and researchers cannot agree on it today. Even though there were still thousands of patients in hospitals and asylums in the 1960s, the medical literature and research on the illness had virtually come to a standstill in 1935.[52]

In the 1920s chronic manganese poisoning was thought by many to be the culprit, because it appears like parkinsonism and can be confused with schizophrenia. Von Economo was adamant that the "intellectual integrity, the absence of delusions and the patients' comprehension of their own pathological condition decide the issue." The Viennese patients were still intelligent (when they could be awakened or brought to sensibility) and realized something was abnormal, but they described their symptoms as compulsive. They felt they were forced to perform the odd and uncomfortable behaviors. Von Economo used this characteristic to identify encephalitis lethargica against any other organic problem. Early on, he thought the epidemic was similar to poliomyelitis, lyssa, or the Borna type of encephalitis that affects horses. Doctors and the public initially suspected botulism poisoning. After all, the world had been wracked with a European war, and the eating of canned food had increased. But no evidence was found to support this theory, either in canned food samples or in the patients' medical tests. Polio was another suspected link; a polio epidemic occurred in New York in 1916, and the disease was widespread at the time. There were many similarities between the diseases, but again, medical proof was lacking. The two diseases possessed too many differences.[53]

The severe abdominal pains that accompanied encephalitis lethargica could be mistaken for appendicitis, because along with the sharp pain comes a rigidity of the abdominal muscles, leading physicians to suspect a swollen appendix.[54] No doubt numerous appendectomies were performed in the 1920s and 1930s because of what was actually a symptom of encephalitis lethargica.

The disease has been identified with influenza because the two diseases often occur in epidemical form at nearly the same time; encephalitis lethargica has been found to precede influenza by one year or to occur simultaneously with epidemical influenza. Epidemics of "sleeping sickness" historically have occurred after or before many of the greater influenza epidemics such as nona, Sydenham's epidemic of the 1600s, and the 1918 flu. Postmortem examination of patients in the 1920s revealed that their lungs were not affected by encephalitis lethargica; patients had exhibited no pulmonary symptoms or lung complications—both are nearly always present with influenza.[55]

In the 1980s Rai Ravenholt, an epidemiologist with the Centers for Disease Control in Atlanta, studied one location where influenza did not occur in 1918—American Samoa. Using death records from nearby Western Samoa (where eight thousand perished from influenza) as well as Seattle archives, he showed that where the influenza virus had not penetrated, neither had encephalitis lethargica. Using this information, he postulated a connection between the influenza virus and encephalitis lethargica—neither of which has been seen again in the virulent form evidenced early in the century. Not all scientists agreed with his finding that both diseases were caused by a single viral agent.[56]

Oliver Sacks felt it was "probable, that the influenza epidemic in some way paved the way for the encephalitis epidemic, and that the influenza virus potentiated the effects of the encephalitis virus, or lowered resistance to it in a catastrophic way." Certainly the fact that between 1918 and 1919, when half the world's population was affected by influenza and more than 21 million people died from it, encephalitis lethargica assumed its most virulent form is evidence of some form of relationship, even if not cause and effect. During late winter, when encephalitis lethargica usually appears, people also experience more colds, which may point to another relationship between influenza and encephalitis. Either they are related in some way, or the encephalitis strikes those whose immunological systems are weakened by colds and flu. The one aspect that confuses the relationship between influenza and encephalitis is the means of transmission: encephalitis lethargica does not spread from person to person, as colds and influenza do. And approximately half the patients hospitalized with encephalitis lethargica in the early twentieth-century epidemic had not earlier had the flu.[57]

The way the 1916–1930 pandemic spread suggests an intermediate transmitter of infection, because there was no proof of direct contamination and because many people never developed it.[58] Encephalitis, whether in epidemic or sporadic form, is usually viral, and can result from a variety of causes: the viruses of HIV, polio, rabies, herpes simplex, herpes zoster, equine encephalitis, and St. Louis encephalitis. Bacteria (typhus), rickettsia (Rocky Mountain spotted fever), and protozoa (toxoplasmosis) also cause a widespread inflammation of the central nervous system, resulting in encephalitis. Arboviruses are a cause of viral encephalitis. They spread in the bellies of blood-feeding insects (arthropods) who carry it from vertebrate host animals to humans. A variety of host animals are known to carry the virus: rodents, birds, pigs, and particularly horses. Mosquitoes or ticks usually serve as transmitters. The virus originates in wild birds and is transmitted to domestic animals by insects who feed on the birds. Domestic animals become unwitting hosts for the disease, which is then passed to humans by mosquitoes who feed on both domestic animals and the occasional human. Animals and humans do not ordinarily pass it to each other; a blood-feeding insect (tick or mosquito) must do the job.

The fact that the 1916–1930 encephalitis lethargica epidemic showed up largely in small towns and rural communities, rather than in larger, more congested cities, seems relevant. There were only a few school outbreaks, particularly at boarding schools in England, which were in rural settings. Country doctors in five or six states in the United States observed and accurately described the disease before any in metropolitan areas were aware of it. The editor of the *American Journal of Public Health* believed this may have meant that those living in more populous settings had already developed an immunity from constant contact with "crowd infection," something that was not possible in small communities.[59] While this may be puzzling for epidemiologists, it does seem to point to arboviral sources of infection, because rural areas would have more horses per capita, as well as proximity between humans, horses, and water troughs (for mosquito larvae), and more wild birds in close contact with humans. Conditions in small towns and rural areas in the early twentieth century as well as seventeenth-century New England included the necessary means for arboviral infection: horses, standing water, and mosquitoes. (See Figure 1 for an overview of the relationship between virus, insect, mammal, and man.)

When an epidemic of arboviral encephalitis develops, its spread is controlled by the number of uninfected hosts that are available, the availability of insects to carry the virus, and climatic controls. Temperatures that are too cool or too hot, changes in humidity and light, all have an effect on mosquito breeding. Eventually the virus plays out when mosquitoes go into their winter hibernation or the temperature becomes too cold and standing water freezes. If conditions are not completely right, the virus will not reappear the following spring but may lie dormant in unhatched mosquito eggs for years until conditions are receptive for them to hatch and the virus reappears again, a process known as recrudescence.[60] When it does reappear, the fact that the hosts are all new and have not been infected before (and have not

been able to build a level of immunity) means that the virus
has adequate conditions for recrudescence; it spreads
quickly and a new epidemic ensues.

Are there other possible causes of encephalitis? What
about head lice? Body lice feed off of human blood, and
though medical authorities are quick to dispel fears that
they are disease vectors, Harvard's Laboratory of Public En-
tomology notes that they are sometimes more than an an-
noyance and an embarrassment. Nit-infested children and
adults may experience itching, sleeplessness, a shortened attention span, and depression. Outside of North America,
body lice do transmit pathogens to humans. Moreover, the
Harvard Laboratory advises physicians and health-care professionals that "a few people remain convinced that their infestation is real, even though they have been examined by
several competent specialists who can find no physical cause
for their discomfort. Some of those patients have become a
danger to themselves and those around them by resorting to
the use of toxic or flammable substances in attempts to rid
themselves of their real or perceived infestation. Such a person may, indeed, be delusional, and should be treated with
care and respect when referred for counseling." Interestingly, in North America and Europe, children are more
likely to be infested with head lice than adults, and Cau-
casians more frequently than other ethnic groups.[61] Elemen-
tary school teachers are well aware of the need to examine
blond-haired children for head lice nits far more thoroughly
than Asian or African-American children, on whom they are
rarely found. Lice as we find them today do not explain a
worldwide pandemic, as happened in 1918, but because
pathogens mutate, perhaps they were partly to blame.

Arbovirus Transmission Cycles

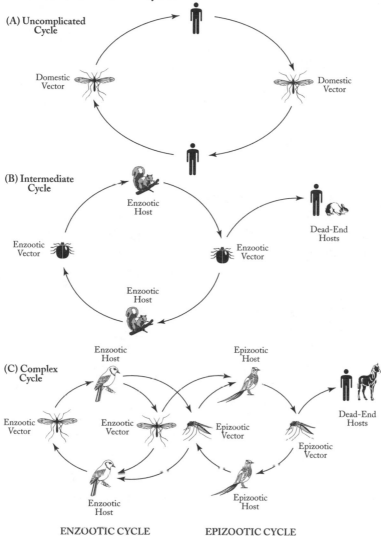

(A) Uncomplicated Cycle

Domestic Vector

Domestic Vector

(B) Intermediate Cycle

Enzootic Host

Enzootic Vector

Enzootic Vector

Dead-End Hosts

Enzootic Host

(C) Complex Cycle

Enzootic Host

Epizootic Host

Enzootic Vector

Enzootic Vector

Epizootic Vector

Dead-End Hosts

Epizootic Vector

Enzootic Host

Epizootic Host

ENZOOTIC CYCLE   EPIZOOTIC CYCLE

Adapted from Monath, *The Arboviruses.*

Figure 1: (A), at top, is an uncomplicated cycle involving only a single vertebrate host and a single primary vector species. In *(B)*, an intermediate cycle, the virus is maintained in an enzootic (constantly present) cycle; humans or other vertebrates are involved only tangentially, usually as dead-end hosts. *(C)* is a complex cycle involving both en- and epizootic (attacking many animals at the same time) subcycles, with different vectors and vertebrate hosts in each subcycle; humans and domestic animals become involved tangentially from the epizootic portion of the cycle.

One puzzling aspect was never considered by physicians working in the large cities. While people were succumbing to waves of influenza and encephalitis lethargica, domestic livestock was experiencing severe mortality as well. In 1932 researchers at the University of California Agricultural Experiment Station filtered a "germ" they thought to be of the virus type, which was causing a highly contagious disease that had been killing off horses and mules in Western states for several years. In one season alone it struck six thousand horses in California. The isolated virus attacked only the brain and spinal cord of the animals, so they named it encephalomyelitis.

Symptoms were similar to the encephalitis lethargica seen in humans. The disease occurred in two distinct types: the sleepy type, where the animal drowsed or went into convulsions; and the walking type, in which the animal paced around and around the field, unable to rest. Sometimes the animal fell down and was unable to get up or roll over. Initially the affected animals could be recognized by an inability to follow when led, a slight wobbly gait, a lack of spirit, a failure to come when called, or, in unbroken colts, a failure to run when approached.

But this was not new. The disease had been common for some time in the Midwest, known there as the Kansas-Nebraska horse plague. While horses and mules were the only animals that usually came down with it, laboratory efforts showed that the virus could be reproduced in monkeys, rabbits, rats, and mice when simply dropped into their nostrils. Researchers were confident that the disease was limited to domestic animals and that it had not progressed into the wild. The only treatment was prevention, and stock owners were advised to isolate infected animals, keep them from drinking in streams, canals, ditches, and ponds, and provide abundant well water for them. The epidemic in California seemed to follow a pattern, the same one as at Salem: animals became infected in the spring, and by autumn, with the cooler weather, the symptoms disappeared.[62]

Dr. Karl Meyer, of the University of California, urged attendees at the meeting of the American College of Physicians in 1932 to examine the brain of every fatal human case of encephalitis they encountered; he had proof that two men had acquired the disease from infected horses, and one had died from it. Realizing that it could, indeed, be transmitted between animals and humans, he continued research, hoping to develop a serum that could confer immunity. He did discover that many horses that had even a slight bout with the disease seemed to have acquired an immunity to greater infection.

Because horses were no longer the economic mainstay they had been, as automobiles and mechanized farm equipment replaced them, the commercial losses were considered unalarming. Three thousand dead horses were processed in one season at a tallow works in California in the 1930s; we may assume that many thousands were being slaughtered or processed due to incapacitating disease. What happened to the products that were shipped around the country? Horse-hair-stuffed sofas and mattresses, soap, candles, glue, even horse meat would have been marketed regionally, perhaps even nationally. Whether these products were able to carry disease to humans is doubtful, but the size and scope of the horse and mule mortality should be reconsidered. Epizootics (epidemics affecting animals) are common but have not been studied as they affect humans. Reconsidering zoonoses, or diseases that can be transmitted from animals to humans, might provide the answer.

The horse plague of the early twentieth century was probably what is now called Western equine encephalitis, or WEE, an infection that occurs primarily in equines (horses and mules) and man. During the 1930s and 1940s WEE was a major veterinary problem in many Western states in the United States, causing thousands of animal deaths. Severe cases of human encephalitis were also reported during that time. The most extensive human outbreak occurred in 1941 in the United States and Canada, when it affected thousands of individuals and caused a large number of mortalities. WEE has also been found in Argentina. Today WEE is found mostly in the Western states along the Pacific Coast.[63]

WEE virus has a wide range of host animals; it has been recovered naturally from squirrels, deer, pigs, and birds. Experimentally it has been found in calves and dogs, but horses are the most susceptible. Young horses are most frequently sickened; older horses appear to have survived the disease with an element of immunity. The disease incubates in the horse for between a few days to two weeks, and the onset of symptoms is abrupt, just as in humans. The horse exhibits hypersensitivity and experiences tremors, neck rigidity, and spasms. Progressive paralysis and death follow within two to five days. Death rates are about 30 percent.

When the disease was isolated in humans in 1938, the incubation period ranged from four to twenty-one days. The onset was sudden, with headache, chills, fever, nausea, and vomiting. Pain and neck stiffness, as well as insomnia, confusion, drowsiness, and tremor ensued. Blindness, speech defects, and paralysis on one side of the body often followed. The death rate was about 10 percent. WEE is most prevalent during the summer and is predominantly a rural disease. Rarely found in urban areas, it was widespread (and still is) in areas where agricultural irrigation is practiced, such as the San Joaquin Valley in California and the Yakima Valley in Washington State. According to the Centers for Disease Control, mosquitoes serve as transmitters that pass the virus between mammals and birds. WEE virus has also been found in tiny parasitic mites that live on wild birds. Although the birds are never sick from the virus, they carry it along their north-south seasonal flight path. Horses and humans are infected with the disease by mosquitoes, but they do not pass it on.

Because mosquitoes are the transmitters, mosquito control has effectively limited the spread of WEE. Insecticides and use of mosquito-larva-eating fish in standing water have helped. So has the use of television (which keeps children indoors in the evening) and window screens. Preventatives help limit mosquito infestations, but they have not eliminated the problem, which continues to be endemic in the very same areas after sixty years of scientific research and control. Although horses are inoculated yearly with a vaccine that prevents the disease, humans are not. There is no treatment for WEE, in either horses or humans, and no cure exists.

An even deadlier strain of equine encephalitis exists on the eastern seaboard of the United States. Called eastern equine encephalitis, EEE is an enzootic that extends from New England to Mexico. It has also been found in Canada, Panama, Mexico, Cuba, Brazil, and the Dominican Republic. Epizootics of EEE in horses have been accompanied by outbreaks of encephalitis in humans. In southern Louisiana in the summer of 1947, fourteen thousand horses and fifteen people were affected at the same time. Coincidentally, in the Dominican Republic in 1948 and 1949 another incidence occurred that affected humans, horses, mules, and donkeys.

First recognized in 1931, EEE is similar to the WEE virus, but the eastern strain is much more contagious to many more animals. Horses, goats, calves, pigs, dogs, and sheep are all susceptible, along with a variety of bird species, particularly pigeons and pheasants. Symptoms in horses are similar to the WEE pattern, along with progressive incoordination, rocking motions, convulsions, and prostration. Horses die or recover, with or without symptoms. The EEE type may also occur in a milder form wherein the horse exhibits a short-term fever with no other symptoms. Death rates are higher than in WEE, running at around 90 percent.

In the late 1930s the disease was found in an outbreak in Massachusetts where the incidence ran very high in children and was severe and fatal. Children under ten years of age accounted for 70 percent of the cases, and the death rate was 65 to 70 percent. Onset was sudden, with fever, restlessness, drowsiness, malaise, stiff neck and back, and convulsions. Death occurred three to eight days after symptoms appeared. EEE appears in summer, along with mosquitoes, which are believed to transfer the disease to humans and horses. The virus is maintained naturally in the wild bird population and only incidentally reaches humans and other mammals. Pheasants farmed commercially are highly infected and have been shown to transmit the disease to humans without the intervention of a mosquito transmitter.[64]

Another strain of equine encephalitis, called Venezuelan equine encephalitis, has been studied since the 1930s. Similar to WEE and EEE, it is significant because it has been shown to pass from an infected horse to a healthy horse through intranasal infection. Horses shed the virus from nose, mouth, eyes, urine, and milk, and can infect by contact. These findings indicate that VEE does not need an insect to transmit the virus between mammals. Experiments with the VEE virus showed that humans were more susceptible to the virus than were laboratory animals.[65]

In the summers of 1917 and 1918 a mysterious epidemic of encephalitis that affected only humans appeared in Australia and reappeared in 1922 and 1925. At first it was called the Australian "X" disease, but it is now thought to have been Murray Valley encephalitis, which has not been found in vertebrates other than humans. It appeared again in 1951 in Australia and New Guinea. The 1951 outbreak resembled a strain that had appeared in Japan, the JE, which caused mental impairment and paralysis with a very high death rate. In Japan, antibodies were found in the blood of horses, pigs, and cattle. Outbreaks in Japan affected humans, horses, and swine. Spread by mosquitoes, the JE virus has been found throughout Asia and was a problem for troops during World War II.[66]

One contemporary caveat about WEE, EEE, and the various other bird-mosquito-mammal encephalitides is that they are cheaply and easily grown, in hens' eggs as well as in the brains of mice. As an agent in biochemical warfare, they would be easy to make, and intensely toxic if spread in aerosol form.

Not all encephalitides are caused by equine-related viruses; ticks can also carry the disease. Tick-borne encephalitis was recognized in Europe and the western Soviet Union after World War II. At first Russian scientists thought it was related to the well-known Russian spring-summer encephalitis, already recognized as a cause of severe paralysis in humans living in forested areas of the Asiatic USSR. Symptoms include an incubation period of eight to eighteen days, with abrupt onset followed by fever, severe headache, nausea, vomiting, and neck pain. Paralysis of the limbs is common; patients are tired and somnolent (sleepy) and may have epileptic convulsions. Some have difficulty breathing and swallowing. When the respiratory system is infected, the patient dies, usually four to seven days after the onset. Recovery can take up to two months, and aftereffects include paralysis and atrophy of the neck and chest muscles. It is a seasonal disease, appearing in April and peaking in June. Usually males are affected. The disease is carried by ticks that live on forest animals, particularly rodents and birds. The disease appeared in a geographical band across forested areas of the eastern Soviet Union, from the northern Siberian to the Russian territory, affecting mostly persons working in virgin forests as loggers and road builders.[67]

Tick bites were also responsible for the spread of a central European meningo-encephalitis that had several distinct features: its highest incidence occurred from April to October, mostly among young adults in rural areas, with no noted spread from person to person, and, in the great majority, only one case per family. This sounds similar to the incidents at Salem, Massachusetts. In Yugoslavia the disease occurred mostly among peasants and villagers, and particularly those who worked as woodsmen and their families. Tick bites were found to have occurred in most of the cases, yet in other European countries, especially the Soviet Union, the drinking of milk from goats or cows bitten by ticks in pastures was one route of infection. For those bitten by ticks, only one per household was infected. For those infected through milk, the whole household became infected. In 1951 an outbreak of milk-borne encephalitis spread to 660 cases in one town in Czechoslovakia. This route of infection can be eliminated by boiling milk before drinking.[68]

As the virgin forests of Europe declined, the tick-transmitted disease was confined to a smaller and smaller geographic area. If this were responsible for disease during the Middle Ages, when forests were being cut down for both fuel and to clear land for agriculture, could the rise in mental disorders that accompanied that period be explained by tick-borne encephalitides?

CHAPTER SIX

What Happened at Salem?

/\

> When champagne was developed in the seven-
> teenth century it was called the "Devil's Wine"
> because no one knew how the bubbles came to
> be, therefore they were assumed to be the work
> of the Devil.
> —John G. Howells, *World History of Psychiatry*[1]

Historical explanations of witchcraft dwell on what Thomas
Szasz calls the "scapegoat theory of witchcraft," which ex-
plores who was accused and why in the context of larger societal issues. Inevitably they fail to examine the accusers or
the "afflicted," who themselves were often tried for witchcraft.

Sociologists have pointed to community-based socioeconomic problems as the causative agent in the events at
Salem. They propose that there were really two Salems:
Salem Town (a prosperous sector on the well-developed east
side of town) and Salem Village (a less-developed, very
swampy and rocky area on the west side). Likening Salem
Village to a troubled backwater, the accusations and afflictions emanated from the west side, where the residents di-
rected their animosity toward their wealthier, more powerful

eastern counterparts by accusing individuals on the east side of witchcraft. Examining the struggles, failures, broken dreams, and lost hopes of the Salem Village residents, sociologists began to view the village as "an inner city on a hill." Social conflict, in this case between prosperous merchants and struggling subsistence farmers, was examined. In the case of the Salem witch hunts, the theory may better explain who was accused and convicted of witchcraft than why individuals were afflicted. Division along class and religious lines has been well documented in determining criminal accusations.[2]

Other investigators have blamed the situation on village factionalism, claiming that Salem Village was rife with suspicious, disgruntled, jealous settlers whose frustrations had festered for years before exploding in the court record with witchcraft accusations and trials. But that does not explain why twenty-two *other* towns in New England were eventually connected to the proceedings in some way; villagers throughout Maine, New Hampshire, Connecticut, Massachusetts, and Rhode Island were brought into the trial records. Victims, accused witches, and witnesses came from other locales as far away as the Maine frontier. Other locations, such as Connecticut, conducted witch trials that preceded or coincided with those at Salem. Choosing to view the problems as power struggles or personality differences within a small village strikes one as too parochial. Many of the possessed claimants barely knew the people they named as their tormenters, in fact several had never even met the persons they accused of fostering their problems—hardly enough tension to support the idea that the entire uproar was based on long-standing animosities. Socioeconomic divisions did engender problems in the region, and while they ultimately may be used to explain who was accused and why, they do not explain the many physical symptoms or who experienced them.[3]

Carol Karlsen has viewed what happened at Salem in her book *The Devil in the Shape of a Woman*, which relates the events to women's oppressed status within Puritan society. She considers New Englanders' "possession" to have been a cultural performance—a ritual—performed by girls, interpreted by ministers, and observed by an audience as a dramatic event. Karlsen claims the possessed individuals exhibited learned behavior patterns and that words and actions varied only slightly among them. The affected women experienced an inner conflict which was explained by ministers as a struggle between good and evil: God versus Satan. The outcome revolved around whether or not the young women would later lead virtuous lives or fall into sin. Karlsen suggests that a woman's possession was the result of her indecision or ambivalence about choosing the sort of woman she wanted to be. She views the possession as a "collective phenomenon" among women in Connecticut between 1662 and 1663, and in Massachusetts from 1692 to 1693. It was a "ritual expression of Puritan belief and New England's gender arrangements," and a challenge to society. It was ultimately a simple power struggle between women and their oppressors.[4]

As to the physical symptoms: the fits, trances, and paralyzed limbs, among others, Karlsen attributes them to the afflicted girls' actual fear of witches as well as the idea that once they fell into an afflicted state they were free to express unacceptable feelings without reprisal. The swollen throats, extended tongues, and eyes frozen in peripheral stares were manifestations of the inner rage they felt toward society; they were so upset they literally *couldn't* speak. Their paralysis was based on anger over having to work; their inability to walk meant they could not perform their expected labor—in other words, a passive-aggressive response to a situation that incensed them. Karlsen views witchcraft possession in New England as a rebellion against gender and class powers: a psychopathology rooted in female anger.[5]

Misogyny may well explain who was accused of witchcraft, but it lacks an explanation for the wide-ranging symptoms, the ages of the afflicted, and the patterns of symptoms that occurred across time and distance in seventeenth-century New England. Scholars who take this route, however, conveniently ignore the fact that men too were accused, tried, and hanged for witchcraft, both in the colonies and in Europe. In fact, Robin Briggs states that though "every serious historical account recognizes that large numbers of men were accused and executed on similar charges, this fact has never really penetrated to become part of the general knowledge on the subject." His research shows that a misogynistic view of witch-hunts lacks complete credibility.[6]

Many researchers have proposed that mass hysteria affected the young women of Salem.[7] The term *hysteria*, essentially a female complaint, has recently been dropped from use by the psychiatric profession in favor of "conversion symptom," which describes the manner in which neurotic patients suffer emotional stress brought on by an unconscious source. This stress or tension can undergo "conversion" and reveal itself in a variety of physical ailments. Conversion, a very pliable disorder, can be explained by almost any societal pressure in any particular culture. It is a psychological catchall for unexplained neurological or emotional problems. But its victims are always the same, according to analysts: unstable females.

Jean-Martin Charcot, a French physician, worked extensively with epileptic and hysteric female patients at the Salpêtrière Hospital in Paris between 1862 and 1870. He laid the groundwork for hysteria theory, calling it hystero-epilepsy. He accused his patients of being deceitful, clever actresses who delighted in fooling the male physician. Charcot's medical students claimed to be able to transfer diseases from hysterics with the use of magnets, something they called the "metal cure." Eventually his professional standing as a neurologist diminished and faded, and he turned to faith healing. Sigmund Freud, one of his students, began his work under Charcot's direction.[8]

A more modern version of the hysteria complex is called Mass Psychogenic Illness, or MPI, which is defined as the contagious spread of behavior within a group of individuals where one person serves as the catalyst or "starter" and the others imitate the behavior. Used to describe situations where mass illness breaks out in the school or workplace, it is usually connected to a toxic agent—real or imagined—in a less than satisfactory institutional or factory setting. MPI is the sufferer's response to overwhelming life and work stress. It relies on the individual's identification with the index case (the first one to get sick, in effect the "leader") and willingness to succumb to the same illness. A classical outbreak of MPI involves a group of segregated young females in a noisy, crowded, high-intensity setting. It is most common in Southeast Asian factories crowded with young female workers; adults are not usually affected. Symptoms appear, spread, and subside rapidly (usually over one day). Physical manifestations usually include fainting, malaise,

convulsions with hyperventilation, and excitement. Transmission is by sight or sound brought about by a triggering
factor which affects members of the group, who share some
degree of unconscious fantasies. A phenomenon more related to the industrial world of the nineteenth and twentieth
centuries than to pastoral village life in colonial New England, MPI does not address the question of why men and
young children, who would not have identified emotionally
and psychologically with a group of young girls, suffered.
The New England colonists scarcely fit the pattern for this
illness theory that demands large groups of people of similar
age, sex, and personality assembled in one confined
location.[9]

Salem's witches cannot, of course, escape Freudian cri-
tique. Beyond the hysteria hypothesis, John Demos, in *Entertaining Satan*, looked at the evidence from the perspective
of modern psychoanalysis. He pointed out that witchcraft
explained and excused people's mistakes or incompe-
tence—a failure or mistake blamed on witches allowed a
cathartic cleansing of personal responsibility. Witches served
a purpose; deviant people served as models to the rest of society to exemplify socially unacceptable behavior. But
Demos's explanation that witch-hunts were an integral part
of social experience, something that bound the community
together—sort of a public works project—does not address
the physical symptoms of the sufferers.[10]

For the most part, examinations of the afflicted individuals at Salem have focused on the young women, essentially
placing the blame on them instead of exploring an organic
cause for their behaviors. Freudian explanations for the
goings-on have attributed the activities of the possessed
girls to a quest for attention. Their physical manifestations
of illness have been explained as being conversion symp-

toms due to intrapsychic conflict. Their physical expression
of psychological conflict is a compromise between unacceptable impulses and the mind's attempt to ignore them.
Demos uses the example of Elizabeth Knapp, whose fits became increasingly severe while strangers gathered to view
her behavior. Instead of considering that she was beset by
an uncontrollable series of convulsions which were likely
worsened by the excited witnesses who refused to leave her
alone, he attributes her worsening condition to her exhibi-
tionist tendencies, motivated by strong dependency needs.
Elizabeth's writhing on the floor in a fetal position is seen as
an oral dependency left over from childhood, causing her re-
gression to infancy.*

   But "inner conflict" simply does not explain the events
at Salem. Neither does the idea that the young afflicted girls
were motivated by an erotic attraction to church ministers
who were called in to determine whether Satan was involved. The girls' repressed adolescent sexual wishes (one
girl was only eleven years old) and their seeking a replace-
ment for absent father figures scarcely explains the toll the
disease was taking on victims of both sexes and all ages. No
Freudian stone has been left unturned by scholars; even the
"genetic reconstruction" of Elizabeth Knapp's past points
out that her childhood was filled with unmet needs, her
mother's frustration because of an inability to bear additional children, and her father's reputation as a suspected
adulterer. "Narcissistic depletion," "psychological transfer-
ence," "a tendency to fragment which was temporarily neu-
tralized"—the psycho-lingo just about stumbles over itself

---

*Von Economo noted that fatigue exacerbated illness and delayed re-
covery. The observers were no doubt making her even sicker, and less
likely to recover. Demos, *Entertaining Satan*, 55.

in attempts to explain the afflicted girls at Salem. But unanswered questions remain: Why the sharp pains in extremities? The hallucinations? The hyperactivity? The periods of calm between sessions of convulsions? Why did other residents swear in court that they had seen marks appear on the arms of the afflicted?[11]

The opinion that the victims were creating their own fits as challenges to authority and quests for fame has shaped most interpretations of what happened in 1692.[12] But would the colonists have strived for public notice and attention? If the afflicted individuals were behaving unusually to garner public notice, why? Did women and men of that era really crave public attention, or would it have put them in awkward, critical, and socially unacceptable situations? How socially redeeming would writhing on the ground "like a hog" and emitting strange noises, "barking like a dog," or "bleating like a calf" be for a destitute young servant girl who hoped to marry above her station? It is difficult to accept that these spectacles, which horrified viewers as well as the participants themselves, were actually a positive experience for the young women. That sort of suspicious activity usually met with social stigma, shunning, or, at the least, brutal whipping from father or master.

Puberty, a time of inner turmoil, is thought to have contributed to the victims acting out through fits, convulsions, and erratic behavior. The victims' inability to eat is explained away as a disorder related to the youthful struggle for individuality: anorexia nervosa. What about the young men who reported symptoms? Freudian interpretation attributes their behavior to rebellion against controlling fathers. How have psycho-social interpretations explained the reason witch trials ended after 1692 in Salem? As communities grew into larger urban units, people no longer knew their neighbors, grudges receded in importance as a factor in social control, and witches were no longer valuable to society. John Demos observes that witchcraft never appeared in cities, and that it lasted longest in villages far removed from urban influence. That linkage between witchcraft outbreaks and agricultural villages is important when establishing a connection with outbreaks of encephalitis lethargica, which appeared largely in small towns and rural areas in the early twentieth century. Rather than accepting the idea that witchcraft receded because it was no longer useful in a community context, one must examine why epidemics occurred in waves and how particular diseases affected isolated population groups.[13]

The situation in seventeenth-century New England fails psycho-social explanation because too many questions remain unanswered. Not only can we not make a strong case that infantilism, sexual repression, and a struggle for individuality caused the turmoil in Salem, but a psycho-social explanation does not answer why the symptoms, which were so *obviously physical*, appeared with such force and then, in the autumn of 1692, largely disappeared from Salem.

Because the complexity of psychological and social factors connected with interpreting witchcraft is so absorbing, the existence of a physical pathology behind the events at Salem has long been overlooked. Linnda Caporeal, a graduate student in psychology, proposed that ergot, a fungus that appears on rye crops, caused the hallucinogenic poisoning in Salem. Her article appeared in 1976 in *Science* while Americans were trying to understand the LSD drug phenomenon. Hers is one of the few attempts made to link the puzzling occurrences at Salem with biological evidence.[14]

Ergot was identified by a French scientist in 1676, in an explanation of the relation between ergotized rye and bread poisoning. It is a fungus that contains several potent pharmacologic agents, the ergot alkaloids. One of these alkaloids is lysergic acid amide, which has ten percent of the activity of LSD (lysergic acid diethylamide). This sort of substance causes convulsions or gangrenous deterioration of the extremities. Caporeal proposed that an ergot infestation in the Salem area might explain the convulsions attributed to witchcraft. If grain crops had been infected with ergot fungus during the 1692 rainy season and later stored away, the fungus might have grown in the storage area and spread to the entire crop. When it was distributed randomly among friends and villagers, they would have become affected by the poisoned grain.

Caporeal's innovative thinking was challenged by psychologists Nicholas Spanos and Jack Gottlieb, who were quick to point out that her theory did not explain why, if food poisoning were to blame, families who ate from the same source of grain were not affected. And infants were afflicted who may not have been eating bread grains. Historically, epidemics of ergotism have appeared in areas where there was a severe vitamin A deficiency in the diet. Salem residents had plenty of milk and seafood available; they certainly did not suffer from vitamin A deficiency. Ergotism also involves extensive vomiting and diarrhea, symptoms not found in the Salem cases. A hearty appetite, almost ravenous, follows ergotism; in New England the afflicted wasted away from either an inability to eat or a lack of interest in it. The sudden onset of the Salem symptoms in late winter and early spring would be hard to trace to months of eating contaminated grain. Ergot was never seriously considered as the cause of problems at Salem, even by the colonists themselves who knew what ergotism was (it had been identified sixteen years earlier) and were trying desperately to discover the source of their problems.[15]

An explanation that satisfies many of the unanswered questions about the events at Salem is that the symptoms reported by the afflicted New Englanders and their families in the seventeenth century were the result of an unrecognized epidemic of encephalitis. Comparisons may be made between the afflictions reported at Salem (as well as the rest of seventeenth-century New England) and the encephalitis lethargica pandemic of the early twentieth century. This partial list, created from the literature, reveals how similar the two epidemics were, in spite of the variation in medical terms of the day:

1692 SALEM	1916–1930s ENCEPHALITIS EPIDEMIC
fits	convulsions
spectral visions	hallucinations
mental "distraction"	psychoses
pinching, pricking	myoclonus of small muscle bundles on skin surface
"bites"	erythmata on skin surface, capillary hemorrhaging
eyes twisted	oculogyric crises: gaze fixed upward, downward, or to the side
inability to walk	paresis: partial paralysis
neck twisted	torticollis: spasm of neck muscles forces head to one side, spasms affect trunk and neck
repeating nonsense words	palilalia: repetition of one's own words

In both times, most of the afflicted were young women or children; the children were hit hardest, several dying in their cradles from violent fits. The afflictions appeared in late winter and early spring and receded with the heat of summer, similar to the way arboviral encephalitis is spread (see the graphs in the Appendix). Von Economo noted that most encephalitis lethargica epidemics had historically shown the greatest number of acute cases occurring in the first quarter of the year, from midwinter to the beginning of spring. The "pricking and pinching" repeated so often in the court records at Salem can be explained by the way patients' skin surfaces exhibited twitches—quick, short, fluttering sequences of contractions of muscle bundles. Cold temperatures cause them to increase in number and spread over the body. Twitches were seldom absent in cases of hyperkinetic encephalitis lethargica during the 1920s epidemic. The skin surface also exhibited a peculiar disturbance in which red areas appeared due to dilation and congestion of the capillaries. Red marks that bleed through the skin's surface would explain the many references in court documents to suspected bites made by witches.[16]

Examining the colonists' complaints in the trial papers uncovers many other symptom similarities: inability to walk, terrifying hallucinations, sore throat, or choking—the list goes on and on.

It is one thing to focus on a relatively small area of New England and to identify an epidemic of encephalitis as the cause of witchcraft allegations and worries. The argument will not hold up, however, unless we can also explain the situation in the witch-ridden areas of Europe. After all, the New Englanders had come from Britain and Europe, where witchcraft had been an entrenched problem. We must examine what actually happened there in order to discover the many similarities. Comparing the records and facts of the two regions, encephalitis lethargica once again makes itself mysteriously present.

Encephalitis lethargica appears in the record in an elusive manner, from Hippocrates' description of a pregnant woman who fell into a coma-like sleep on through the oral tradition that developed in the sixteenth and seventeenth centuries with allusions to individuals awakening from a deep sleep (most recognizable is the tale of Sleeping Beauty, a young woman who sleeps for years, then reawakens and continues on with her life). What may be called the "case of Barbara Kremers" proves that encephalitis lethargica was already endemic in parts of Europe at the time of the miseries in Salem, and was perhaps the cause of some (or much) of the epidemical mental illness that plagued the Continent in the sixteenth century. In 1573 Johann Weyer, the Paris-educated physician, accompanied Duke William of Julich, Berg, and Cleves, to Königsberg, where he heard the locals talking about the miracles happening all around the area. One topic of conversation at royal dinner tables was that certain people were able to live without eating. (Whether this was intriguing simply because the affliction had potential to minimize economic support for the peasantry is not known.) The scientific imagination was stimulated by stories of young women and girls who were able to go for long periods without food of any kind, yet still lived. It was said that in the nearby town of Unna there was just such a girl. In fact there were stories of "fasting" women all over the area, but Weyer found the home of the ten-year-old girl in Unna.

Barbara Kremers was well known, in fact a celebrity in the area. Her mother told Weyer that the miracle of her not eating had come after the girl had suffered a severe illness that lasted six weeks. After that she had remained mute for six months. She had come out of her stupor, but since recovering she had eaten and drunk nothing, nor did she urinate or move her bowels. The parents were extremely proud of their daughter; she was "good" and devout. The nobility and learned men of the area had already visited her, and the Unna city council had even presented her with a signed certificate testifying to the wonder. People streamed into the cottage daily to see Barbara and left money for her and her family. When Weyer saw Barbara she looked well enough but was unable to walk without crutches, due to being "lame." She also had a twist in her arms and was unable to straighten them.[17]

Obviously she had the same symptoms that would occur less than 120 years later in Salem. But in Unna her affliction was interpreted as a miracle rather than diabolical. Weyer, the skeptic, invited her to his home where he treated her medically. When she recovered her ability to talk and eat, he accused her of being a fraud. No one punished Barbara or her parents, but the city council took back the certificate, and life resumed its ordinary ways in Unna.[18] Again, the interpretation of the afflicted person's state made all the difference in how she was treated and what was expected of those around her.

Barbara's experience led Weyer to assume that one aspect of witchcraft belief had been misinterpreted due to physical causes: the incubus. The term appears often in European and New England witchcraft complaints. It was a term that described an oppressive, nightmarish burden that pushed down on a person while in bed. Sometimes it was interpreted as an evil spirit that descended upon and had sexual intercourse with women as they slept. Johann Weyer claimed that the incubus was not a devil but a common ill-

ness that appeared at night. He could not completely identify it, but, he said, "Those who are tormented by it think in
their sleep that a heavy burden is pressing on them, this in-
terfered with their breathing and consequently with their
power to give voice and speak. They force themselves to cry
out but cannot. They have terrifying dreams and they imag-
ine that someone has come to do them harm." He was describing the very same symptoms so many had attested to in
Salem, such as John Louder, Richard Coman (he claimed an
invisible person had sat on his chest), and Phoebe Chandler.
It was the same feeling as those experienced by sufferers in
the 1920s who felt controlled by an opposing force. "Patients so affected find that as soon as they 'will' or intend or
attempt a movement, a 'counter-will' or 'resistance' rises up
to meet them," Oliver Sacks writes. "They find themselves
embattled, even immobilized, in a form of physiological
conflict—force against counter-force, will against counter-
will, command against countermand." The terrifying dreams
that Weyer identified were the same that had caused Ann
Putnam to see a man burning on a spit and Mercy Lewis to
see coffins and winding sheets waiting for her.[19]

Strange occurrences were not unknown to sixteenth-
century England, where victims succumbed to a peculiar
sort of unexplained sudden death, popularly called "being
taken by a planet." In 1593, a plague year, one town saw
eighteen fatalities caused by it (seventeen of them were
children). Victims were said to be "planet struck" or "moon
struck"—in the belief that the victim had been suddenly
overtaken by the hostile influence of a planet. The afflicted
person sat in a trancelike state for some time before dying.
Nicholas Culpeper, a seventeenth-century physician, described this unusual catalepsy: "in English it is called conge-
lation, or taking, and by the ignorant struck with a Planet. It

is the sudden detention and taking both of body and mind, both sense and moving being lost, the sick remaining in the same figure of body wherein he was taken; whether he sit or lie, or whether his mouth and eyes were open or shut, as they are taken in the disease so they remaine." What then was considered to be epilepsy, accompanied by stomach cramps and unconsciousness, was called the falling-sickness, because people fell to the ground in fits, or lunatism, because they were under the moon's influence. Being planet-struck was very similar to experiencing encephalitis lethargica; perhaps it was a forerunner of the epidemic strain of the virus.[20]

Life expectancy for those who survived infancy in Salem in the late 1600s was far better than it had been in Tudor or Stuart England. Several men and women were in their seventies and eighties at the time of the 1692 trials. New Englanders lived in small, scattered villages and settlements which isolated them from the sort of epidemics that had prevailed throughout England. In the large towns and cities of Europe, infectious diseases regularly killed 10 percent or more of the population between 1485 and 1610. In England by the late sixteenth century the very poor had fallen below the level of subsistence into actual starvation. Their immunity levels were lowered, their susceptibility to disease increased. They had become the perfect population for epidemical disease.

Only a small fraction of Europe was concerned with witchcraft. The major cities and government centers were never involved; the witch persecutions were largely common to rural areas. Europe was a literary society, though, and no matter how removed a community might have been, written materials were available nearly everywhere that described the activities of both the accused witches and their prosecution. Even in North America, Cotton Mather had read about the possessed children of Sweden, and he noted that the children's stories he heard from the afflicted children in Boston and Salem were much the same as the European accounts.[21]

A difference between Europe and New England was the relatively common report that the person accused of being a witch was subject to trancelike states or fits. In the Salem records the accused did not suffer from such states, only the accusers did. In Europe many of those tried for witchcraft evidently were afflicted themselves.[22]

One young English girl, Mary Glover, had an experience similar to sufferers in New England—obviously a case of encephalitis lethargica which in Europe and Britain was as unfathomable as ever. She fell into a "grievous fit," which daily increased in strength and strangeness. She evinced incredible body contortions during the fits, turning backward like a "hoop" with her head near her hips. She rolled and tumbled about in this manner so violently that she had to be held to prevent bashing her head on the bedstead and floor. Handlers noted that she was "all over cold and stiff as a frozen thing."[23]

Cases in southern Lorraine around 1609 reported other symptoms of encephalitis lethargica, such as the woman who fell sick with her mouth twisted grotesquely aside; another thought she was bewitched because she felt several rat-sized animals running around inside her body and became bedridden with great pains in her limbs. Several others claimed to have pains in throats and arms. People felt sudden weights on their chests at night so that they could not breathe, reported seeing other persons in the room, and feared they would be physically attacked by them. In Lancashire around 1612, people complained about weight on their chests while in bed at night, an inability to speak, hallucinations about vicious cats and catlike animals, panic attacks, partial paralysis, and illnesses lasting a year or more—all were chronicled similarly to those at Salem, and similar to the epidemic of the 1920s.[24]

Villagers in early modern Europe, as well as in New England, were not ignorant, superstitious half-wits who attributed all their problems and troubles to witches. They were invariably facing something they simply could not control, let alone recognize. In fact, probably many cases of afflictions were never brought into the records because there was simply no one to accuse credibly. One powerful motive for accusing people was the hope that they could effect a cure. Certainly in New England the records are filled with descriptions of how a supposed witch was asked to touch an afflicted person, or to explain why they caused the trouble, or to identify how to cure the person. It was when the supposed witch was unable to remove the purported curse that things became muddled. Determining who was accused of bewitching the sufferers and why they were singled out is a matter of the socio-dynamics of the people and their situation. *Why* the problems erupted is a question of organic causes.

Domestic animals were at the forefront of the European witch persecutions, just as they were in New England. Animals in both societies represented a large investment and were extremely valuable to their owners. When animals were affected with unusual deaths, and people suffered economic losses, these complaints entered the court records too. In Norfolk, England, in the early seventeenth century, several animals were reported to have been affected with sudden attacks and death, such as the calf which was "taken so strangley as if the back were broken, and much swollen

and within the space of three or four days it died," followed
by another that ran in circles and also acted as if its back
were broken.[25]

As in New England, the animals were butchered and examined. A horse that died strangely was found to have flesh
that was blackened as if it were burned; a milk cow in St.
Croix in 1604 was reported to have been "all burned" in the
organs, and its flesh was black. Many animals were described as having died "as if rabid," though they had not actually *been* rabid. In 1598 a farmer in Lorraine had lost
animals every year for the preceding nine years. Some died
as if rabid, others with their muzzles rotting off, as if they
were "burned." One of the farmer's pigs did survive, left
with a "twisted neck."[26]

Scholars have puzzled over why the witchcraft persecutions in Europe were limited to a relatively short period of
time—a 150-year span—and limited to the general region of
the Rhine (but extending into Sweden, eastern England,
and Scotland for short periods). The witch persecutions occurred in two distinct waves (see Figure 2). The first, between the late 1300s and the mid-1400s, were located in a
few isolated Alpine valleys. They were off the main trade
routes, but the people were not backward by any means.
The area had a highly developed economy and social institutions. The second wave of witch persecutions occurred
between 1590 and 1640, when most of the trials were held.[27]
The persecutions took place in a relatively defined area
along the Rhine River, extending 50 to 150 miles on each
side, and running from Switzerland to the Spanish Netherlands. Half of all the known trials occurred in that general
area. Mysteriously, a large area of Europe never experienced
witch trials during this period. In those areas that were involved in the witchcraft persecutions, intense activities

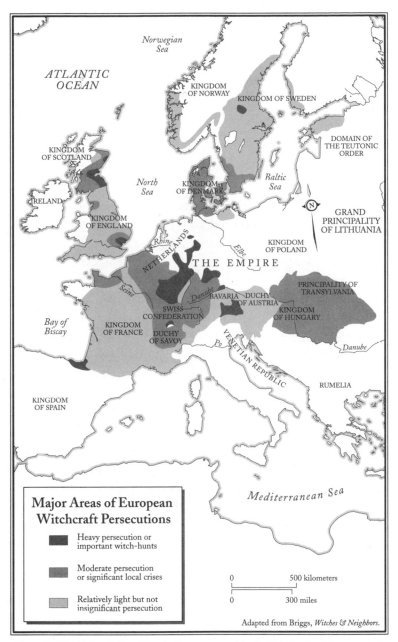

Figure 2: *Clearly the witch problems in Europe centered in the Rhineland, with some activity in Sweden and along the east coast of England and Scotland.*

lasted only a few years. Why did most places escape the trauma?

Limited to time and distance, explanations have focused on social, religious, economic, and political tensions. The Thirty Years' War must have played a part, most writers note, but Briggs points out that the witchcraft persecutions took place in areas that were hardly touched by that conflict.[28]

One explanation bears further study, if such can be done from a distance of more than three hundred years. Witch persecutions in Europe were centered in western France, some in eastern England, Austria, and Switzerland, as well as northeast Italy. There were few elsewhere. Looking at a map of the flight path of migratory birds (see Figure 3), we see how closely they match up. Birds migrating from sub-equatorial western Africa fly directly over these areas as they head north each spring. Traveling all the way to the northern parts of Sweden and Russia, air currents take them directly over the very areas of witchcraft worries. A second route, over Austria and Switzerland, heads south directly over Italy and straight to the West African coast at the Gulf of Guinea.[29] Migratory birds may have brought disease from western Africa to Europe, where a virus in their blood was extracted by arboviral mosquitoes who then fed on peasants and villagers. Just as viral encephalitis is spread by mosquitoes today, a unique form may have been spread by warm-blooded wild birds who passed it on to mammals and humans. Wild birds are recognized as carriers of viral and bacterial disease; in fact the Centers for Disease Control monitor flocks of chickens, known as sentinel chickens, throughout the United States, testing the domestic birds' blood weekly for evidence of virus. They are the alert

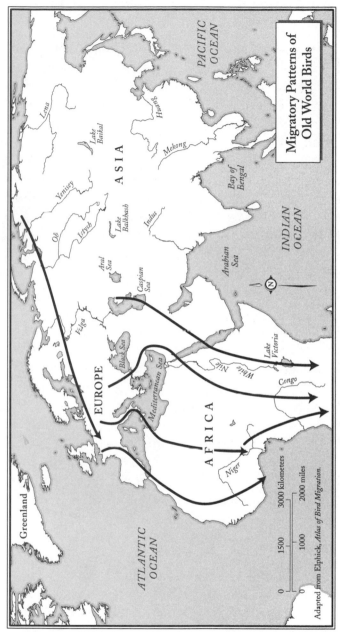
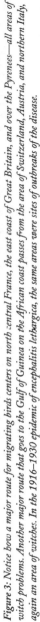

Figure 3: Notice how a major route for migrating birds centers on north-central France, the east coast of Great Britain, and over the Pyrenees—all areas of witch problems. Another major route that goes to the Gulf of Guinea on the African coast passes from the area of Switzerland, Austria, and northern Italy, again an area of witches. In the 1916–1930 epidemic of encephalitis lethargica, the same areas were sites of outbreaks of the disease.

mechanism by which epidemiologists determine that an epidemic has developed.

Of course one must extrapolate this theory in order to answer the mysteries of Salem. How would the same birds have flown to New England—an incredulous impossibility? They didn't. Migratory birds on the east coast of North America travel north from the Guiana highlands and Venezuelan river bottoms, passing through Barbados and the West Indies on their way to Massachusetts and points north (see Figure 4). The popular birding areas of Martha's Vineyard and Plum Island (which is very near Salem) are havens for viewing the migratory flocks of ducks, gulls, and swallows today.[30] In the past, huge flocks of pigeons would have darkened the sky too. Just as malaria was brought to Barbados by the slave ships that plied the Atlantic, so too encephalitis lethargica could have come. Malaria, spread by the *Aedes aegypti* mosquito, is believed to have been transferred by malaria-ridden female mosquitoes who laid eggs in water barrels and buckets aboard seventeenth-century ships.[31] Today wild birds harbor numerous viruses, and at times they can be transferred to domestic birds, livestock, and even people. Perhaps this unique strain of encephalitis is still around, passed from bird to bird by mosquitoes or tiny mites, without ever moving into a different species. At some point, when conditions are suitable, it will be forced to expand into other hosts.

If the cause of encephalitis lethargica was to be found in Europe (where the epidemic of the 1920s began), English immigrants to North America could well have brought it in their own bodies as immune carriers, or if it was spread by domestic animals, in the bodies of the livestock they brought with them to the colonies, which would have served

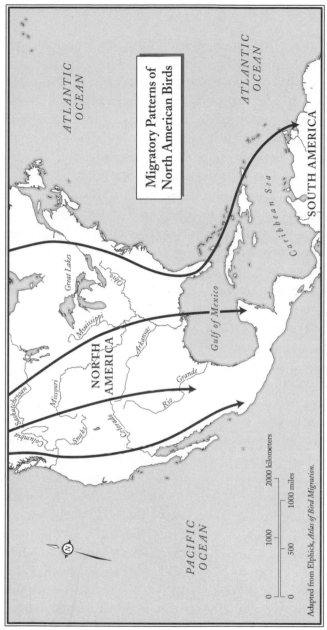

Figure 4: Note the migration path that passes directly over New England, through the West Indies, en route to Venezuela and Guiana. Points that host large populations of birds during migratory periods include Plum Island and Martha's Vineyard, Massachusetts, as well as the Delaware Bay and Chesapeake Bay regions. About 330 species migrate to South America for the winter, including swallows, thrushes, tanagers, and shorebirds. Hurricanes or violent storms may force the birds to alter their regular route.

as repositories of microbes waiting for mosquitoes or other insects to pass them to the people.

But encephalitis is not known to pass person-to-person. Encephalitis spreads in several ways, but usually an insect carries it from an infected animal. If carried by an arboviral insect, as most encephalitis epidemics are today, how could an insect have brought the virus to New England? On the very ships that brought goods and people from the nearest sea-trading destination: the West Indies. Mosquitoes that carried malaria are known to have come by ship from Africa to the West Indies, then northward. Mosquitoes have a life span of only a few weeks, but they bred and flourished in the water kegs every ship carried. By the time a ship made it to Boston or Salem, mosquitoes could have infected all the passengers and their livestock.

If so, why wasn't Boston hit harder than Salem? Perhaps the terrain and geography of Salem help explain that question. Looking at a map of the Salem area, one sees that those who complained of bewitchment or affliction (or those who complained for their sick relatives) lived on the west side, nearer to the forest, and in rocky, swampy terrain—a river flowed directly through the area (see Figure 5). The geography and climate in the Massachusetts and Connecticut colonies were very favorable for mosquitoes. John Josselyn, a visitor to the colonies in the early seventeenth century, wrote in his travel diary that "The country is strangely incommodated with flyes, which the English call Muskataes, they are like our gnats, they will sting so fiercely in summer as to make the faces of the English swell'd and scabby, as if the small pox for the first year." Josselyn's description of the terrain indicates the potential for heavy mosquito infestation. He wrote that the town of Ipswich (directly north of Salem on the coast) was "situated by a fair

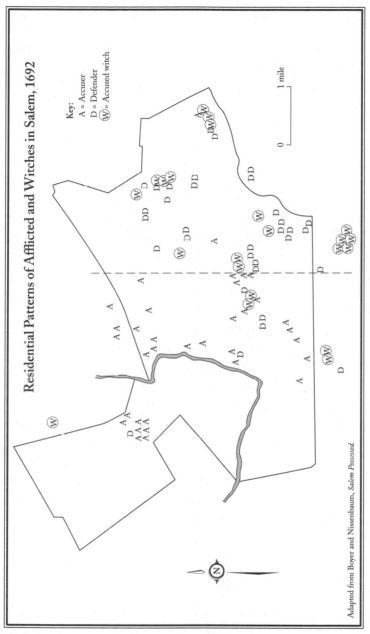

Residential Patterns of Afflicted and Witches in Salem, 1692

Key:
A = Accuser
D = Defender
(W) = Accused witch

0 |————| 1 mile

Adapted from Boyer and Nissenbaum, *Salem Possessed.*

Figure 5

River, whose first rise is from a Lake or Pond twenty mile
up, betaking its course through a hideous Swamp for many
miles." Hampton, another nearby settlement, had "a great
store of salt marshes and Cattle, the land is fertil, but full of
swamps." Sudbury, too, had a "great store of fresh marshes,
and Arable land, and they have many Cattle, it lyeth low, by
reason whereof it is much indamaged with flouds." He describes the area around New-Berry as "wide venting
streams," and Wenham as being "very well watered . . . well
stored with cattle." Near Linn "runs a great Creek into a
great marsh called *Rumney-marsh*, which is four miles long,
and a mile broad." In the seventeenth century the region
suffered from poor drainage and swampy conditions: a mosquito larva's paradise. And every village would have had a
water-powered grist mill or sawmill and a mill pond perfect
for mosquito breeding.[32]

Today New England suffers from EEE, which is passed
from horses to humans through mosquitoes. Although the
type of EEE present today seldom appears in epidemic
form, it could have mutated over the past three hundred
years. Because the colonists lived on the edge of the "wild,"
it would have been easy for mosquitoes to pass the virus
from a wild animal to a domestic animal, and from there
carry it from the infected domestic animal to other domestic
livestock and humans. The virus could then spread through
a settlement or urban area exponentially. This sort of transmission is usually seen in villages close to the site where the
arbovirus is active. It results in a sporadic, endemic epidemic (one that lies within the community for a long time),
which can become an enzootic, an epidemic that also affects
animals. If the wild-animal host population disappears because of changes in the ecosystem (hunting, overgrazing by

domestic animals, drought), the insects turn to humans and domestic livestock to feed, increasing the epidemic. Changes wrought on the environment by colonial settlement may have forced the live mosquito population to feed on the humans and their animals in order to survive.[33]

Many arboviruses exist in certain ecological niches and are restricted to a particular geographic region by climate, lack of a source of host vertebrates, or an inability to travel to other areas. That may explain why the epidemic did not spread far beyond the New England area. With travel by sea rather than land in the mid to late 1600s, few residents and their domestic animals would have traveled overland to colonial settlements beyond New England. The climate was favorable in the region, with an unusually warm and wet summer in 1691, followed in 1692 by an early spring and dry summer. This fits the weather pattern conducive to most epidemics of arboviral encephalitis in North America today.[34]

Larger weather patterns for the area may be part of the equation, too. Recent research on eight-hundred-year-old bald cypress tree rings reveals a devastating drought in the Virginia colony area between 1587 and 1589, which may explain the disappearance of the colonists of Roanoke Island. Although the tree rings are not from the same geographic area, perhaps they can be used if only to recognize what may have been general weather patterns for the east coast of the continent (see Figure 6). The tree-ring data show that the summers between 1680 and 1683 were the wettest in more than a century. That period was followed by a drought in the late 1680s, then a trend from 1690 to about 1698 of alternately very wet and abnormally dry summers. If we apply these findings to New England, the wet conditions were perfect for mosquito infestations, followed by droughts when so many larvae hatched that the adults were forced to feed on humans and livestock.[35]

Human behavior also increases susceptibility to arboviruses. Young unmarried women and girls would have been given tasks such as milking cows and gathering eggs (early morning and evening tasks that occur just when mosquito activity increases), tending plants in the garden (mosquitoes are frequently found in grasses and vegetation; a garden would have fostered them), and going for water from the well, which meant water in buckets and barrels that would have attracted egg-laying mosquitoes. The increased exposure to mosquito bites would explain why so many young women were affected; the disease is also more prevalent among young adults. Evidently the colonists recognized that their problems were not being caused by contagious disease passed from person-to-person contact, because they did not quarantine the afflicted as was the practice for those suffering from smallpox or other identifiable communicable diseases of the day.[36]

Ultimately the witch-hunts—or at least the complaints of afflictions—ended in Salem in the autumn of 1692, and there were no more complaints the following year. An arboviral encephalitis epidemic would have receded in the fall, when the air and water grew too cold for mosquitoes' survival. By the time spring arrived, the situation had altered, and the epidemic appeared to fade. Encephalitis epidemics, like many other contagious epidemics, often recede for years—sometimes decades—between recrudescence periods. Either the agents mutate and disappear to return years later, or they run out of susceptible hosts—the only ones left are those who have an immunity to the infection.

Ticks too might have been to blame. Just as in the

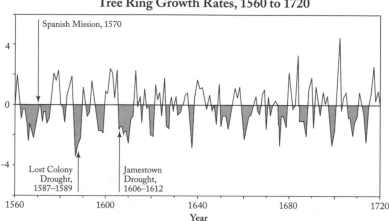

Adapted from David W. Stahle, Malcom K. Cleaveland, Dennis B. Blanton, Matthew D. Therrell, and David A. Gay, "The Lost Colony and Jamestown Droughts," *Science*, April 24, 1998.

Figure 6: This graph represents the amount of moisture available during the summer growing season, as shown in tree ring samples. Recordings above the center line indicate more moisture than average; those below the center line indicate drier than usual conditions.

spread of tick-borne encephalitis throughout the northern region of Russia, ticks played a part in spreading the disease across the virgin forests of temperate North America. Peasants who worked in the forest as woodcutters were affected in Russia during the epidemic of the 1950s; in Salem, in the seventeenth century, residents also worked as woodcutters and loggers. The Putnam family, in particular, were engaged in logging and woodcutting (and in fact were involved in arguments over whether they were taking logs from property they did not own). If the Putnams brought ticks bearing disease into their homes on their bodies or clothing, other members might have been affected. Reverend Parris's household could have been infected from the large amount of firewood he negotiated to supply his family, as part of his

salary. Because they were his strongest supporters, the Putnams would likely have been the ones to cut and deliver the
wood to his doorstep. Firewood, in the form of large logs
used in colonial fireplaces, might have harbored wood ticks
that had gone into winter hibernation but came out of the
bark when logs were stored beside the hearth in a warm
New England house. Infestations of ticks and body lice
were common in colonial homes where laundry could not be
done during the winter (nowhere to dry the wet clothing)
and baths were rarely taken.

Another disease that results in encephalitis is endemic to
the New England area even today. Lyme disease is a contemporary problem in New England, and there is little reason to think that it would have been absent from the area in
colonial times. It is an infectious disease caused by bacteria
spread by deer ticks. Both people and animals can be infected with Lyme disease. It is a serious but not fatal disease
today. Found throughout the United States, it is most common along the East Coast, the Great Lakes, and the Pacific
Northwest. In Massachusetts, deer ticks are most often
found along the coast and are common in the Connecticut
River Valley. The disease most likely spreads between late
May and early autumn, when ticks are active. So tiny that
the larvae are no bigger than a pencil point, the ticks live for
two years, during which they can infect wild and domestic
animals as well as people.

Symptoms of Lyme disease include a rash where the tick
was attached—which may appear anywhere between three
days and a month after the innocuous bite. Some times the
rash looks like a small red doughnut. Other signs include
itching, hives, swollen eyelids, and flulike symptoms such as
fever, headache, stiff neck, sore muscles, fatigue, sore throat,
and swollen glands. The symptoms go away after a few

weeks, but without medical treatment nearly half the infected people will experience the rash again in other places on their bodies. In the later stages, three major areas—the joints, the nervous system, and the heart—may be affected even months after the tick bite. People with Lyme disease can develop late-stage symptoms even if they have never had the rash. About 10 to 20 percent of the people who do not get treatment develop nervous system problems: severe headache, stiff neck, facial paralysis, or cranial nerve palsies, and weakness and/or pain in their hands, arms, feet, or legs. Symptoms may last for weeks, often shifting from mild to severe and back again.

These symptoms are found in the present form of Lyme disease; the disease could likely have mutated over the centuries, because hallucinations and paranoia, along with lethargy, are not found in today's tick-borne version of Lyme disease. Questions and problems arise when connecting Lyme disease to the situations in 1692 or 1920, but it is another factor to consider. Could ticks have been common in Salem? The colonists did not bathe regularly, and they lived close by their domestic animals. Ticks could have wintered inside the home, carried in on firewood. They would have found ample hiding places in the seams of the heavy woolen clothing commonly worn by the colonists.

What about 1920? A common nuisance of that era was the "bedbug," chinch bug, or *Cimex lectularius*. Jar lids filled with arsenic were placed under bedsteads to keep the critters from climbing into bed and feeding on people's blood. Head lice have been common throughout the ages; today's rampant epidemics in schools are nothing to ignore, though scientists reassure us that neither bedbugs nor head lice carry any type of disease. Perhaps they did at one time. Many avenues must be explored, much research must be

done. Perhaps we will never know what caused encephalitis
lethargica.[37]

Physical symptoms can be readily compared, but what about
the "invisible world," as Cotton Mather called it? What
about the terrifying visions of the Black Man or, alternately,
the pleasant visions of the White Spirit? What about the
brilliant light so often reported? The colonists were dumbfounded by the awed or fearful recitations of people who
saw the apparitions, and were forced to address the
question: should invisible entities be admitted in court? In
Salem, visions ("spectral evidence") were ultimately
allowed in court. Other courts in New Hampshire, Connecticut, and other Massachusetts towns did not allow them.
Because a court of law is a place where damages and complaints are lodged, only the visions that portended destruction and harm were examined. No one examined the
brilliant light or White Spirit seen in some nighttime hallucinations. Whether they were disregarded because they could
not be interpreted or understood, or whether they were
ignored because they seemed to point to madness, it is
difficult to ascertain.

CHAPTER SEVEN

Alternative Outcomes

/\\

These are the Lord's doings and truly marvelous
in Mine Eyes!
—Reverend Nicholas Gilman, diary,
January 1742

Could events in Salem have turned out differently? In
some ways they did, in an adjoining area a generation later.
Reading *The Diary of Nicholas Gilman*, we see how similar
symptoms were interpreted far differently than they had
been at Salem. Nicholas Gilman (1708–1748) was a nonde-
script New Hampshire minister who was of the next generation of religious men in New England. A historian who
studied his small and incomplete diary in 1919 called it "a
distressing record of the growing mental unbalance of a gentle but naturally melancholy nature."[1]

Although Nicholas Gilman was a Harvard graduate and
colonial minister, he was unlike Parris or Mather. When his
flock began experiencing spectral visions, he chose to interpret the apparitions far differently. When Elias Parkman was
"Siezd with a Paralytic Fit Suddenly in the Street," Gilman
remarked, "Man Knoweth not his Time." He kept busy
reading books about treating measles, smallpox, and diphtheria in order to advise and treat his family and congrega-

tion. When his sister had "Hysteric disorders" following measles, he dosed her with warm honey, rum, and oil, with a drink of hyssop afterward. He gathered wood betony (Bishopswort) in the forest because it came in handy as a folk cure for "falling-sickness," colds, and convulsions. It had also been used as a protection against evil spirits in England, was good for "fearful visions," and could drive away devils and despair. It was the English colonial equivalent of Prozac.[2]

In 1740 sickness was all around Exeter, New Hampshire, and many died. People were "struck speechless" in the street and died from what was said to be "Apoplectick" fits. Young Mary Reed began telling Gilman about her visions of the "Holy Spirit." That fall he visited his brother's house, where during the night there were "3 Women taken with Hysterick fits—Mrs. Jerusha Lord, filled with Light and Comfort—I was Calld out of my Bed before Morning to see her. Found her in a Very pleasant joyfull frame."[3]

As winter drew on, individuals began to tell Gilman they had been "awakened." In December, "Between Meetings a great crying out, among people in Anguish of Spirits. Many awakened." Gilman himself was ill with fever and sore throat; two of his children died of diphtheria that winter, but he recovered, and by January his preaching had become more emotional. He began to lead all-night services, "Sometimes Praying, then Singing, Exhorting, advising, and directing, and Rejoycing in the Lord."[4] He reported that the congregation would begin screaming if he left for a break. His diary began to show large sections written, then scratched out. He admitted to having trouble concentrating.[5]

During one of his nightlong January services, a man "Saw a White Dove come down into the Meeting house," and "Two Angells." A young woman also said she had seen angels. Another man saw a "bright light like an exceeding bright star" come down from the ceiling and settle on a beam. Things accelerated, and there seemed to be so many visions and apparitions in the community that Gilman noted, "the circumstances are too many to record." His interpretations were entirely different from those of Parris and Mather—"these are the Lord's doings and truly marvelous in Mine Eyes!" he wrote with enthusiasm.[6]

While preaching he watched as youths "under Good Influences and of regular Life" began to call out and "fall into Visions." In response, Gilman read his notes about various church members' trances and visions to the enthralled congregation, capping the evening with "the Eighth Chapter of the Revelations." He even allowed a youth to take the pulpit, "when He in a Vision warnd and exhorted them most Earnestly."[7]

In 1742 diphtheria raged through the area, and the death rate in New Hampshire towns soared: "There is one Relative gathered into the Grave after another," Gilman sadly wrote. He was busy with funeral after funeral. As if in response, revivalism and itinerant preaching increased as rapidly as the disease. New Light preachers frequently met at Gilman's home and discussed the topic of "visions," then they all spread throughout the region preaching in the lively style of the popular English revivalist George Whitefield. The Great Awakening had begun.[8]

By the winter of 1743 the excitement had faded for Gilman; his spirits were low, people's interest had passed. The revival was winding down. In January he settled down to read a book by a young woman in Boston about her "Miraculous Recovery."[9] Perhaps he was thinking it would bring back the glorious hours he had experienced when his congregation was "awakening" all around him. The title says it all: *Address to Young People, or Warning to Them from One Among Them, Yet May Be Called Warning From the Dead; given by Mercy Wheeler of Plainfield, a Person Confined to a Bed of Languishing For More Than Five Years.*

Mercy Wheeler had been born in Plainfield, Connecticut, in 1706, the fourth daughter in a family of three sons and ten daughters. Her amanuensis, Samuel Stearns, also of Plainfield, and probably the local minister, described the situation in the foreword he wrote to "her" book:

> She was a Child very remarkable for Health and Agillity of Body, until the Twentieth Year of her Age; and then was taken sick of the burning Ague in the latter end of the summer of 1726, and hath never since enjoyed one Days health; she was in the Summer following, brought so low, as that she was wholly Confined to her Bed, (which Confinement continues to this Day) and reduced to such exceedingly low Circumstances, as that every Day, for many Days and Months together, was expected by all about her would have been her last. Her strength was so intirely gone as she could not move her Head, Hand or Foot so as to help her self in the least; and her Speech wholly failed; so that she has not been able to speak, whispering so as to be heard by any, but those of the Family, much used to her; and they guessing more by the motion of her Lips than by any Voice which could be heard.
>
> Her Flesh is much consumed. Her Appetite to Food is so gone, that it is admirable to think how she has been sustained; her digestive Faculty so weak, that she has not been able to Eat either Bread or Flesh of any sort, for the space of Five Years and more: supported only (as to outward means) by some thin, liquid Food, put into her Mouth with a Spoon; who will say but that the Power and Goodness of

God is very remarkable in sustaining her so long time by such slender and inconsiderable Means.

Mercy's condition sounds very similar to that of Barbara Kremers, the young European girl studied by Weyer in 1573. But the city councilmen of Plainfield did not recognize her with a certificate as those in Unna had done for Kremers. Instead they sought remedy for her problems through the church:

> No Care or Cost, has been wanting on her behalf; Physicians far and near, have been consulted and improved (to great Expence) for her. Nor have her Parents neglected to seek GOD for a Blessing on their Endeavors of this Nature: Diverse Days of Fasting and Prayer have been attended in her Father's House, with the assistance of the Pastor of the Church here, and other Christian Friends, and also the Pastors of divers of the neighboring Churches; and several Sermons adapted to the occasion, have been Preacht in her hearing . . .

In spite of their efforts, the young girl had a difficult recovery:

> She still remains weak and low, but (thanks be to GOD) she is under some revivings. Most remarkably the last *April* it pleased GOD to open her lips and *her mouth doth shew forth his Praise:* she was on a sudden enabled to speak with a Voice so loud as to be heard into another Room in the House, to the surprize of the whole Family; this reviving, and power of Speech she seems greatly affected with, and blessed GOD for the same . . .

Pastor Stearns was quick to utilize the girl's situation to bolster interest (and attendance) at church, pointing out that the girl's troubles were an atonement for the congregation's sins:

> I think indeed she seems to be under the powerful Impressions of the Spirit of GOD, and touch'd with a sense of the prevailing sins of the present Times, laments the decay of the Power of Godliness amongst us.[10]

Mercy undoubtedly was not able to write a single word and perhaps did not even utter any. Her situation gave others, notably Stearns, the chance to promote church attendance and prayer among the young people of the region. The book begins:

> *My dear friends, relatives, and neighbors* who are now in your youth, and come to that Time of Life, which is called Flower of the Age. . . . And now God has given me ability [to speak again], I think it is my duty to you O young People, to address myself to you, by way of Advice, Councel, & Caution: to you which are near to me in the bonds of nature, in an especial manner; yea and to all the whole neighborhood, and whom ever are or shall be acquainted with my circumstances. . . .
>
> Remember! O young people, Remember that Childhood & Youth is vanity. What shall I say to you to persuade you to be serious? How shall I order my speech to you, so as to prevail with you to hearken me now, that so God may hearken to you Another Day?[11]

The warnings increase and continue, quoting the Old Testament and exhorting young people to be saved, followed by urgings to go to church, perform daily prayer, and attend family worship. Not the words of a sickly young woman, the rambling paragraphs resemble pages of notes from a sermon. Obviously Stearns, the "borrowed pen," was doing his utmost to use Mercy's situation to frighten young adults into attending church. It is difficult to believe that the weakened girl actually used such phrasing. Nevertheless, the book reveals that she had suffered a bout of encephalitis lethargica, recovered somewhat, and her affliction was being promoted as a message from God rather than a consequence of witchcraft. Much had changed in Massachusetts in the thirty-four years since Salem's rash of witchcraft cases.

Undoubtedly Stearns, as well as Reverend Gilman and his friends in the itinerant ministry, were not the only ones who had found the random cases of illness useful. In the 1700s colonial society embraced a religious experience filled with visions, talking in tongues, shaking, and seeing "the light." Individuals afflicted with encephalitis were viewed as positive, even visionary, members of the community.

CHAPTER EIGHT

Could Encephalitis Lethargica Return?

All the most acute, most powerful, and most
deadly diseases, and those which are most difficult to be understood by the inexperienced, fall
upon the brain.
 —Hippocrates, "On the Sacred Disease"[1]

The physical symptoms and in particular the psychotic
symptoms that accompany encephalitis lethargica fit the
complaints and symptoms reported in late-seventeenth-
century New England. There would have been no way for
the colonists to know what they were dealing with; it was
not until the invention of the electron microscope in the
1940s that scientists were able to truly understand viral
agents. There would have been no way to heal those who
were affected; encephalitis still has no known prevention or
cure.

Von Economo clearly believed that his work would be
extremely important in later years; he was optimistic that
medical professionals who followed him would add to the
body of knowledge about encephalitis lethargica and its effects. In fact, no one bothered to do much about it after

1935. Has encephalitis lethargica been relegated to history, much like the black death but slightly more mysterious and less threatening? Perhaps not. Encephalitis lethargica afflicted five million persons worldwide, many who were left crippled for decades. Any disease that widespread and virulent certainly cannot simply be forgotten.

In 1983 English pediatricians were surprised to discover that many of the children brought to the Cambridge University Hospital suffering psychiatric and neurological symptoms appeared to have encephalitis lethargica. They were lethargic, heard "voices," had expressionless faces, experienced double vision and weight loss, attempted suicide, and exhibited oculogyric crises of the eye muscles. Writing in *The Lancet,* English physicians supported the idea that encephalitis lethargica still exists, and they suggested it be considered in children with chronic emotional and behavioral disturbances coinciding with atypical depression.[2]

Professor John Oxford, currently researching the encephalitis lethargica epidemic of the 1920s, has been examining brain samples taken from victims who perished. The tissue samples had been preserved in wax and stored in the basement of a London hospital, in hopes that technology could at some point determine what caused the epidemic. Oxford's research has been unable to detect influenza RNA in the samples, but he has continued the search, exhuming seven coal miners from their permafrost graves in the arctic. One survivor of the epidemic, Philip Leather, remains in a London hospital, and upon his death his body may provide additional answers. Doctors are diagnosing new cases of encephalitis lethargica which will enable researchers to perform detailed studies using these patients. As elusive as encephalitis lethargica is, determining a cause for the disease may take more time and better technology than is currently available. Oxford explains, "We are searching for a ubiquitous virus, present in a wave, which then vanished."[3]

Cheyette and Cummings, in the *Journal of Neuropsychiatry*, point out that study of the 1920s epidemic and the symptoms of encephalitis lethargica may provide insights into many contemporary neuropsychiatric disturbances that look eerily like the symptoms of encephalitis lethargica. They draw comparisons between encephalitis lethargica and contemporary problems such as sleep disorders, tardive dyskinesia, Tourette's syndrome, obsessive-compulsive disorder, depressive disorders, severe conduct disorders in children, and paraphilias (deviant sexual behaviors including pedophilia, transvestism, and rape).[4]

In 1930 von Economo predicted that future epidemics might be unrecognizable, perhaps with mild symptoms that would be ignored. He enjoined psychologists to examine the case descriptions of encephalitis lethargica in order to avoid speculation about psychological manifestations "built on sand." He urged "purely speculative psychologists" to consider the findings regarding encephalitis lethargica from a factual, organic basis. "Every psychiatrist who wishes to probe into the phenomenon of disturbed motility and changes of character, the psychological mechanism of mental inaccessibility, of the neuroses, and etc. must be thoroughly acquainted with the experience gathered from encephalitis lethargica," he wrote. "Encephalitis lethargica can scarcely again be forgotten."[5]

Satanic Possession and Christian Beliefs

Even today there is a need to differentiate between demonic possession and mental illness. A book that comes highly recommended by the National Alliance for the Mentally Ill (NAMI) can help devout Christians to reconcile the Bible with neurobiological disorders in family members. *Strength for His People: A Ministry for Families of the Mentally Ill*, written by Pastor Steven Waterhouse, accepts the integral belief of many that demonic influence is rare but possible. Many Christians who encounter schizophrenia and other disorders wonder at some point whether their relatives are possessed. Even those who do not maintain the belief in demonic possession often encounter those who do. A condescending response or dismissal of belief in the supernatural is arrogant as well as ignorant. The Bible does mention demons more than a hundred times. But the Bible makes a distinction between disease and possession, in Mark 6:13, for example.

The Bible formulates at least six factors that aid in distinguishing schizophrenia from demonic possession. Waterhouse describes them as:

1. *An attraction to vs. an aversion to religion.* People with neurobiological disorders are attracted to religion, while demons want nothing to do with it.

2. *Irrational speech vs. rational speech.* According to the New Testament, demons spoke logically and rationally. People in the throes of unmedicated neurobiological disorders will often speak nonsense and jump quickly between unrelated topics. Their verbalizations are sometimes confusing, often nonsensical.

3. *Ordinary learning vs. supernatural knowledge.* In the New Testament, demons spoke through people to convey knowledge. People with neurobiological disorders have no access to knowledge they have not encountered before.

4. *Normal vs. occultic phenomena.* Demonological occult happenings, such as levitation and telepathy, affect others in the room, but a person with a neurobiological disorder is the only one to experience anything odd, such as hearing voices or seeing hallucinations.

5. *The claim to be possessed.* Experienced professionals believe that those who claim to be possessed are very likely *not* possessed. Demons want to be secretive and do not voluntarily admit their presence.

6. *Effects of therapy.* If medicine alleviates the problem, it was not demonic possession. If only prayer solves the problem, it was probably not neurobiological disorder.

CHRONOLOGY

This chronology drawn from the literature may be a resource when comparing patterns of disease in both the colonies and Europe. The individuals named are from the Salem records, unless noted.

400 B.C.	Hippocrates (460–388 B.C.) writes about epidemics of lethargy and mania in ancient Greece.
400 A.D.	"Dancing" epidemics in Europe.[1]
1573	In Germany, Barbara Kremers, a ten-year-old girl afflicted with encephalitis lethargica, treated as a miracle because she does not eat. She recovers, is called a fraud.[2]
1620	Just before English arrival at Plymouth, Indians in the area are killed by plague. An Indian population of thirty thousand dwindles to three hundred.[3]
1646	"Malignant fever" arrives in springtime in Massachusetts.
1647	"Epidemical sickness" kills many.
	First New England witch trial, Windsor, Connecticut, in the spring of 1647.
	Alice Young is hanged.[4]
1648–1649	Smallpox epidemic in Massachusetts; whooping cough epidemic.[5]

Boston harbor quarantines ships from Yucatan, Barbados, and Cuba to keep yellow fever out.

1650s   Influenza strikes Barbados, killing five thousand to six thousand.[6]

1660–1661   Influenza epidemic in Massachusetts, few fatalities.[7]

1661   Heavy spring floods in New England.

1662   Elizabeth Brown suffers from choking, leg pain, "pinching & pricking," stomach pain; two months later she has "frensy" and supernatural distemper. Thirty years later she remains a "sad case" mentally.[8]

1663   Witchcraft outbreak, Springfield on Connecticut River and Hartford, Connecticut.[9]

1665   Influenza epidemic in Massachusetts, few fatalities.[10]

1666   Smallpox outbreak.

1668   Influenza in Connecticut.[11]

1671   In October, Elizabeth Knapp of Groton, sixteen years old, experiences seizures, laughter, falling, leg and throat pain, choking, leaping and running around; barks like a dog; is unable to speak; attempts suicide; has oculogyric crisis during fits; shouts nonsensical words during fits; has remission between fits when she is sensible. Boston racked with "griping of the guts, feaver, & Ague, & bloody Flux."[12]

1672   James Carr has hallucinations.[13]

1672–1673   Encephalitis lethargica ("sleepy sickness") epidemic in London accompanied by hiccups and recognized as the first epidemic of encephalitis lethargica in historical times.[14]

1675	Influenza sweeps Europe and colonies; encephalitis lethargica epidemic in Europe.[15]
	Hurricane hits New England.*
1676	Influenza-like illness in Massachusetts.
1677–1678	Smallpox in Salem.[16]
1678	William Stacy, age 22, experiences hallucinations.
	Samuel Gray, age 28, has hallucinations and sees blinding "light"; his infant dies from "fits."[17]
1678–1679	"Epidemical colds" in Massachusetts.†
1679	John Stiles, a boy in Newbury, shows symptoms of fits, hyperactivity, and "pinching & pricking"; he barks like a dog and attempts to eat sticks, yet is calm between fits.[18]
1680	Rebecca Jacobs becomes insane, is still mentally deranged twelve years later.[19]
1682	The Perley girl of Ipswich is taken with fits which last until she dies three years later.[20]
1683	Hurricane hits New England.
	"Feaver & ague which was very mortal," in Salem and Connecticut.[21]
1684	Malaria in French Canada.
	Jarvis Ring of Salisbury has red marks ("bites") on his skin, is unable to walk or speak, and experiences "pinching & pricking."[22]
1685	Smallpox in Boston, an extremely fatal epidemic.

*Hurricanes are often followed by increased disease epidemics because so much of the terrain is disturbed and water left standing.
†Duffy thought it might have been influenza.

Smallpox in Virginia, on Byrd plantation, comes by slave ship from Gambia.

In Massachusetts, the Shattuck child has fits; is unable to walk; limbs and joints are "twisted"; has oculogyric crisis; is restless and hyperactive. His mental impairment remains twelve years later.

Samuel Perley's daughter dies after three years of fits.

John Louder, age 24, has hallucinations and feels a heavy weight on his chest.[23]

1686–1687   Young mother, Christian Trask, becomes "distracted" and has fits, and exhibits psychotic behavior until she commits suicide.[24]

John Cook also experiences hallucinations.

1687   Abigail Faulkner's husband becomes "distracted"; his symptoms return five years later.[25]

1688   Influenza epidemic in England, Ireland, Virginia.[26]

1688–1689   Measles epidemic hits Massachusetts.

1689   Smallpox appears in Massachusetts in October with a ship from the West Indies and spreads through Boston and surroundings. A public fast and prayer day is held on March 6; the epidemic intensifies during the summer.[27]

John Fuller and Rebecca Shepherd both die strange, violent deaths that relatives attribute to a "malignant fever."[28]

Benjamin Holton has fits and stomach pain, is blinded by light, and dies in March, two days after Rebecca Shepherd.[29]

Dorcas Hoar is sick in bed for some time and has "gone out of the head."[30]

1689–1690 Smallpox epidemic in Salem and rest of Massachusetts, probably brought back from attack on Quebec by local militia.[31]

1690 Priscilla Stacy dies after two weeks of unusual behavior during which she "screaked out" in a strange manner.[32]

1691 In the spring, John Maston's child, of Salem, and Phillip White's child, of Beverly, both die of fits.

1692 Smallpox in Portsmouth and Greenland, New Hampshire; January to end of March, smallpox in Marblehead.

Heavy spring floods in New England.

So many are reported ill in 1692 that only deaths and important cases are mentioned here:

Thomas Green's twin babies die of fits.

James Beale, a boy in Marblehead, dies from choking fits after recovering from smallpox.

Jonathan Putnam's eight-week-old baby dies of fits in April.

Sometime that winter, Mary Warren's mother has "taken ill & dyed."

Abigail Faulkner's husband's fits return. He had first become afflicted in 1687.

Abigail Soames has "laid bedrid a year."

In September the Cole children see a blinding "light" and have fits—they are the last reported cases to be afflicted in 1692, ending the witch craze era.[33]

1693   Yellow fever comes to Boston by ship from Barbados.[34]

1695   In Germany, a physician writes about an isolated case of encephalitis lethargica, with sleep and eye disorders.[35]

1697   Influenza returns to New England after ten years.[36]

1697–1698   Cerebrospinal meningitis in Dover, New Hampshire; could be typhus.[37]

1700   Malaria appears in the Mississippi Valley, probably from fur traders.

1712   Encephalitis lethargica epidemics in Germany.[38]

1720   Colonists begin variolation (inoculation with live matter); already practiced among Africans who came from the Caribbean.

1721   Smallpox arrives in Boston by ship from Barbados.

1735   First diphtheria, "throat distemper," outbreak in North America. Reverend Gilman notes it in New Hampshire.

1745   Sleeping sickness epidemic in France.[39]

1780–1782   Influenza epidemic that includes sleepiness, delirium, convulsions, and neuralgia. Could be encephalitis lethargica.[40]

1830–1833   Influenza epidemic in France that includes mental symptoms.[41]

1889–1891   Encephalitis lethargica cases appear in Europe and the United States.[42]

1890–1891   Nona epidemic of sleeping sickness in Italy and Hungary, probably an encephalitis lethargica epidemic after an influenza epidemic.[43]

1895   Encephalitis lethargica in parts of the United States and Europe.

1916–1930   Encephalitis lethargica pandemic spreads around the world. It appears in several epidemics that recede and return with slightly different symptoms.

1918–1919   Encephalitis lethargica appears in the United States.[44]

1919   Hiccup epidemic in Vienna.[45]

1966–1969   Oliver Sacks begins working with post-encephalitic patients from the 1920s who are still hospitalized at Mt. Carmel Hospital, New York.

STATISTICAL APPENDIX

Persons Reporting Afflictions, 1692

Table 1

Victims of Witchcraft by Age and Sex

		Male	Female	Sex not known	Total
Ages	0–9	8	6	3	17
	10–19	8	12	—	20
	20–39	49	19	—	68
	40–59	10	18	—	28
	60+	1	2	—	3
	Total	76	57	3	136

(From Demos, *Entertaining Satan*)

Table 2

Possessed Accusers by Sex and Age, New England, 1620–1725

Age	Female	Male	Total
Under 10	3	1	4
10–19	27	5	32
20–29	13	1	14
30–39	8	1	9
40–49	3	0	3
50–59	1	0	1
60–69	3	0	3
70 and over	1	0	1
Total	59	8	67

Possessed Accusers by Sex and Marital Status, New England, 1620–1725

Marital Status	Female	Male	Total
Single	42	9	51
Married	20	1	21
Widowed	3	0	3
Total	65	10	75

(From Karlsen, *The Devil in the Shape of a Woman*)

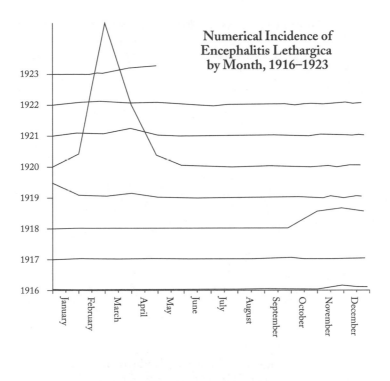

Numerical Incidence of
Encephalitis Lethargica
by Month, 1916–1923

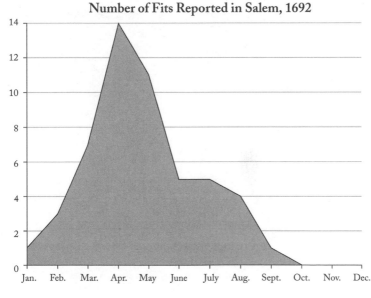

Number of Fits Reported in Salem, 1692

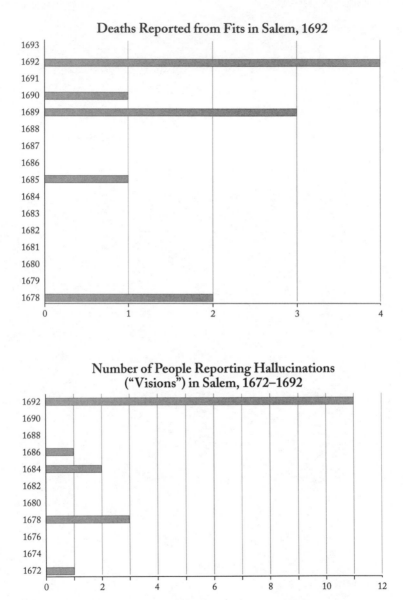

Deaths Reported from Fits in Salem, 1692

Number of People Reporting Hallucinations ("Visions") in Salem, 1672–1692

These were reported in court records during the 1692 hearings. The periods between clusters of incidents follow the pattern for a disease that retreats into a "trough" of inactivity for several years, to revive again.

NOTES

PREFACE

1. John M. Taylor, *The Witchcraft Delusion in Colonial Connecticut, 1647–1697* (New York, 1908), 74.

2. Paul Boyer and Stephen Nissenbaum, eds., *The Salem Witchcraft Papers*, vol. 1 (New York, 1977), 5 (hereafter referred to as *Salem Witchcraft Papers*).

3. Daniel J. Boorstin, *The Americans: The Colonial Experience* (New York, 1958), 21–23.

4. *Salem Witchcraft Papers*, 9.

5. Carol F. Karlsen, *The Devil in the Shape of a Woman: Witchcraft in Colonial New England* (New York, 1987), 262.

6. Robin Briggs, *Witches & Neighbors: The Social and Cultural Context of European Witchcraft* (New York, 1996), 110, 115.

7. Oliver Sacks, *Awakenings* (New York, 1990); Phil Long, "Encephalitis Fears Spread to 11 Counties," *Miami Herald*, August 30, 1997. In 1990 an epidemic of St. Louis encephalitis caused authorities in Florida and Texas to cancel evening baseball games and other events during times of mosquito activity. Laurie Garrett, *The Coming Plague: Newly Emerging Diseases in a World Out of Balance* (New York, 1994), 576. In Sonoma County, California, in 1977, risk for Western equine encephalitis (which affects both horses and humans) was advised. Robert Digitale, "Equine Virus Warning in County," *The Press Democrat*, August 16, 1997, B1–B2. "Disney Takes Precautions Against Encephalitis" (Orlando) *Sun-Sentinel*, August 28, 1997. The theme park was nearly shut down, curtailing golf, fishing, and swimming at dusk each day. In 1990 there had been an outbreak there which resulted in 230 people infected and eleven deaths. Warnings for Lyme disease, another potential source for encephalitis, are ongoing in New England, where it is carried by deer ticks.

CHAPTER 1: THE WITCH CRAZE IN SEVENTEENTH-
CENTURY NEW ENGLAND

1. John Hale, "A Modest Inquiry into the Nature of Witchcraft, 1702," in *Narratives of the Witchcraft Cases, 1648–1706*, George Lincoln Burr, ed. (New York, 1914; reprint, New York, 1968), 400–401 (page citations are to the reprint edition).

2. Carlo Ginzburg, *The Night Battles: Witchcraft and Agrarian Cults in the Sixteenth and Seventeenth Centuries*, trans. John and Anne Fedeschi (Baltimore, 1983), xix.

CHAPTER 2: THE AFFLICTED

1. *Salem Witchcraft Papers*, 99.

2. Hale, "Modest Inquiry," 413, 414.

3. Cotton Mather, "Memorable Providences Relating to Witchcrafts and Possessions" (1689), in Burr, ed., *Narratives*, 101.

4. Ibid., 102.

5. Ibid.

6. Fielding H. Garrison, *The History of Medicine*, 4th ed. (Philadelphia, 1913, reprint 1929), 303; Mather, "Memorable Providences," 103.

7. Mather, "Memorable Providences," 105.

8. Ibid., 105.

9. Ibid., 107, 108.

10. Ibid., 108.

11. Ibid.

12. Ibid., 109.

13. Ibid., 110.

14. Ibid., 119.

15. Ibid., 122.

16. Deodat Lawson, "A Brief and True Narrative of Witchcraft at Salem Village" (1692), in Burr, ed., *Narratives*, 154.

17. Mather, "Memorable Providences," 22.

18. Ibid., 23.

19. John Demos, *Entertaining Satan: Witchcraft and the Culture of Early New England* (New York, 1982), 111.

20. Mather, "Memorable Providences," 27.

21. Demos, *Entertaining Satan*, 135.

22. Ibid., 346; *Salem Witchcraft Papers*, 13.

23. *Salem Witchcraft Papers*, 94, 104, 318, 623, 595.

24. Ibid., 623.

25. Ibid., 99.

26. Ibid., 636.
27. Ibid., 634.
28. Ibid., 216.
29. Ibid., 439.
30. Ibid., 96.
31. Ibid., 192, 99, 563, 376, 604, 743.
32. Ibid., 175.
33. Ibid., 434, 599, 736.
34. Ibid., 563, 435, 372.
35. Ibid., 402; John Josselyn, *John Josselyn, Colonial Traveler,* Paul J. Lindholt, ed. (Hanover, N.H., 1988), 128.
36. *Salem Witchcraft Papers,* 376.
37. Ibid., 559.
38. Ibid.
39. Ibid., 125.
40. Ibid., 94.
41. Ibid.
42. Ibid., 669.
43. Ibid., 372.
44. Ibid., 803.
45. Ibid., 228.
46. Ibid., 420.
47. Ibid., 218.
48. Ibid., 103, 193.
49. Ibid., 194.
50. Ibid., 232.
51. Ibid., 567; Mather, "Wonders of the Invisible World," in Burr, ed., *Narratives,* 238,233.
52. Mather, "Wonders," 240.
53. Ibid., 242, 243, 244.
54. *Salem Witchcraft Papers,* 821, 822.
55. Ibid., 805.

CHAPTER 3: THE RESPONSE

1. Mather, "Memorable Providences," 128.
2. Ibid., 250.
3. Lawson, "Brief and True," 250.
4. Ibid., 155.
5. J. Franklin Jameson, ed. *Narratives of the Indian Wars, 1675–1699,* Original Narratives of Early American History series (New York, 1913), 47.

6. Ann Giudici Fettner, *Viruses: Agents of Change* (New York, 1990), 47; C. E. A. Winslow, Wilson G. Smillie, James A. Doull, John E. Gordon, *The History of American Epidemiology* (St. Louis, 1952), 52.

7. John Duffy, *Epidemics in Colonial America* (Baton Rouge, La., 1953), 22; Frederick F. Cartwright, *Disease and History* (New York, 1972), 123.

8. Duffy, *Epidemics*, 45, 48.

9. Ibid., 194.

10. Ibid., 206.

11. Richard Mead, *A Discourse on the Plague* (London, 1744; reprint New York, 1978), 87.

12. Ibid., 131, 132.

13. Ibid., 128, 162; Robert S. Desowitz, *Who Gave Pinta to the Santa Maria? Torrid Diseases in a Temperate World* (New York, 1997), 67.

14. Winslow, et al., *History of Epidemiology*, 19.

15. Robert Calef, "More Wonders of the Invisible World," in Burr, ed., *Narratives*, 313.

16. Ibid., 314.

17. Ibid., 316, 346.

18. Ibid., 361.

19. Taylor, *Witchcraft Delusion*, 50, 56.

20. Ibid., 68.

21. Ibid., 65, 105, 109.

22. Ibid., 105.

23. Ibid., 106.

24. Frances Hill, *A Delusion of Satan* (New York, 1995), 17.

25. William H. McNeill, *Plagues and Peoples* (New York, 1976; reprint New York, 1993), 47 (citations are to the reprint edition); Jameson, *Indian Wars*, 71, 74.

26. Samuel Parris, *The Sermon Notebook of Samuel Parris, 1689–1694*, James F. Cooper, Jr., and Kenneth P. Minkema, eds. (Boston, 1993), 8, 12.

27. Ibid.,., 117, 118.

28. Ibid., 305, 307.

29. Ibid., 309; Paul Boyer and Stephen Nissenbaum, *Salem Possessed: The Social Origins of Witchcraft* (Cambridge, Mass., 1974), 77.

30. Mather, "Wonders," 209.

31. Ibid., 259.

32. Ibid., 260.

33. Ibid., 265.

34. Ibid., 266, 267.

35. Ibid., 275, 279.

36. Jameson, *Indian Wars*, 75.

37. Ibid., 213; Mather, "Decennium Luctuosum," in Jameson, *Indian Wars*, 209.

38. Ralph Boas and Louise Boas, *Cotton Mather, Keeper of the Puritan Conscience* (New York, 1928), 170, 172.

CHAPTER 4: MENTAL ILLNESS AND THE PERSECUTION OF WITCHES

1. Mather, "Wonders," 249.

2. Thomas S. Szasz, *The Manufacture of Madness* (New York, 1970), 73; John G. Howells, *The World History of Psychiatry* (New York, 1975), 47.

3. Gregory Zilboorg and George W. Henry, *A History of Medical Psychology* (New York, 1941), 133, 143.

4. F. C. Stam, "The Netherlands," in Howells, *World History*, 150; George Mora, "Italy," in Howells, *World History*, 59.

5. Zilboorg and Henry, *History*, 107.

6. Ibid., 139.

7. Ibid., 166.

8. Ibid., 241.

9. Briggs, *Witches & Neighbors*, 8.

10. Zilboorg and Henry, *History*, 233.

11. Ibid., 235.

12. Ibid., 25.

13. Ibid., 275.

14. Ibid., 86.

15. Szasz, *Manufacture*, 87.

16. Ibid., 187.

17. Zilboorg and Henry, *History*, 300.

18. Ibid., 550, 266, 551.

19. Ibid., 551; Garrison, *History of Medicine*, 738.

20. Szasz, *Manufacture*, 73, 74; Zilboorg and Henry, *History*, 260.

CHAPTER 5: THE FORGOTTEN EPIDEMIC

1. Fettner, *Viruses*, 135; S. R. Cheyette and J. L. Cummings, "Encephalitis Lethargica: Lessons for Contemporary Neuropsychiatry," *Journal of Neuropsychiatry and Clinical Neurosciences*, vol. 7, no. 2 (Spring 1995), 125.

2. Sacks, *Awakenings*, 12. Sacks, an English doctor, began working with patients fifty years after they initially came down with encephalitis

lethargica. They exhibited Parkinson's symptoms and had been in a suspended state of sleep since the epidemic. His experimental use of the drug L-dopa to stimulate the patients from their immobile state led to his book *Awakenings*, which was later made into a feature film.

3. Simeon Margolis, ed., *Johns Hopkins Symptoms and Remedies* (New York, 1995), 245. An aside related to viral encephalitis: President Woodrow Wilson, who was sickened during the 1918 influenza epidemic, is also thought to have been a victim of the encephalitis epidemic that was spreading at the time. His personality changes as a result of encephalitis have been postulated to explain his shift in policies during the last two years of his presidency. This theory is cited in *The Throwing Madonna: Essays on the Brain*, by William H. Calvin (New York, 1991), 111.

4. Frederick Tilney and Jubert S. Howe, *Epidemic Encephalitis (Encephalitis Lethargica)*, (New York, 1920), 4.

5. Constantin von Economo, *Encephalitis Lethargica: Its Sequelae and Treatment*, trans. by K. O. Newman (London, 1931), 13, 149; Sacks worked with patients who did not exhibit signs of sickness for several years after their initial infection, or who exhibited very mild signs initially that receded and later reappeared with intensity as parkinsonism.

6. Hippocrates, "On the Sacred Disease," in *Writings* (Franklin Center, Pa., 1979), 346.

7. Hans Zinsser, *Rats, Lice and History* (Boston, 1934, 1963), 115.

8. Cheyette and Cummings, "Encephalitis Lessons," 126. Also in Sacks, *Awakenings*, 320; Fettner, *Viruses*, 132.

9. R. R. Dourmashkin, "What Caused the 1918–30 Epidemic of Encephalitis Lethargica?" *Journal of the Royal Society of Medicine*, vol. 90 (1997), 515; Sacks, *Awakenings*, 321.

10. Dourmashkin, "What Caused?" 515; Cheyette and Cummings, "Encephalitis Lessons," 126, "Sleeping Sickness," *Science*, supplement (March 30, 1923), viii.

11. Cheyette and Cummings, "Encephalitis Lessons," 126; "Sleeping Sickness," viii; von Economo, *Encephalitis Lethargica*, 14.

12. Von Economo, *Encephalitis Lethargica*, 12, 14, 15. Von Economo found the highest incidence in the months of February and March.

13. Tilney and Howe, *Epidemic Encephalitis*, 79.

14. Von Economo, *Encephalitis Lethargica*, 13.

15. Ibid., 27; Sacks, *Awakenings*, 67; von Economo, *Encephalitis Lethargica*, 32, 34.

16. Von Economo, *Encephalitis Lethargica*, 36.

17. Ibid., 37, 38.

18. Ibid., 39.

19. Sacks, *Awakenings*, 389; von Economo, *Encephalitis Lethargica*, 39, 40.

20. Von Economo, *Encephalitis Lethargica*, 39.

21. Ibid., 42.

22. Ibid., 40.

23. Ibid., 42.

24. Ibid., 39.

25. Ibid., 45.

26. Ibid., 47.

27. Ibid., 49, 51.

28. Ibid., 68, 59, 117.

29. Ibid., 118–122.

30. Ibid., 124, 119.

31. Ibid., 129; Fettner, *Viruses*, 135.

32. Von Economo, *Encephalitis Lethargica*, 129.

33. Tilney and Howe, *Epidemic Encephalitis*, 21–23.

34. Ibid., 35–37.

35. Ibid., 43.

36. Ibid., 45, 46.

37. Ibid., 47–49.

38. Ibid., 51, 53.

39. Ibid., 54.

40. Ibid., 72, 82, 112.

41. Von Economo, *Encephalitis Lethargica*, 90.

42. Ibid., 41.

43. Ibid., 53, 54.

44. Ibid., 60, 63, 149; Cheyette and Cummings, "Encephalitis Lessons," 128.

45. Von Economo, *Encephalitis Lethargica*, 63.

46. Cheyette and Cummings, "Encephalitis Lessons," 129.

47. Fettner, *Viruses*, 135.

48. Von Economo, *Encephalitis Lethargica*, 68.

49. Sacks, *Awakenings*, 28.

50. Cheyette and Cummings, "Encephalitis Lessons," 129.

51. Ibid., 129.

52. Sacks, *Awakenings*, xxvii.

53. Von Economo, *Encephalitis Lethargica*, 136, 3.

54. Ibid., 97.

55. Ibid., 98, 88.

56. Fettner, *Viruses*, 138.

57. Sacks, *Awakenings*, 13; Tilney and Howe, *Epidemic Encephalitis*, 72.

58. Dourmashkin, "What Caused?," 516.

59. "Encephalitis Lethargica," *American Journal of Public Health*, vol. 13 (June 1923), 487.

60. Paul Reiter, "Weather, Vector Biology, and Arboviral Recrudescence," in Thomas P. Monath, ed., *The Arboviruses: Epidemiology and Ecology*, vol 1 (Boca Raton, Fla., 1988), 252.

61. Richard J. Pollack, "Head Lice: Information and Frequently Asked Questions," *Head Lice Information*, Harvard School of Public Health, April 3, 1998. Available from http://www.hsph.harvard.edu/headlice.html; Internet (accessed May 22, 1998).

62. "The Cause of the Epidemic Among Horses in the West," *Science*, vol. 75, no. 1928 (February 19, 1932), 10; "Encephalomyelitis," *Science*, supplement, vol. 75, no. 1946 (April 15, 1933), 8.

63. N. Goldblum, "Group A Arthropod-borne Viral Diseases," in *Zoonoses*, J. Van Der Hoeden, ed. (Amsterdam, 1964), 359.

64. Ibid., 365.

65. Ibid., 367.

66. Ibid., 369–372.

67. Ibid., 383.

68. Ibid., 386.

CHAPTER 6: WHAT HAPPENED AT SALEM?

1. Howells, *World History*, xv.

2. Demos, *Entertaining Satan*, 58.

3. Taylor, *The Witchcraft Delusion in Colonial Connecticut*, is a good source about other communities that experienced the same situation. Karlsen, *Devil*, 225.

4. Karlsen, *Devil*, 231, 240.

5. Ibid., 249, 251.

6. Briggs, *Witches and Neighbors*, 259.

7. Ilza Veith, *Hysteria: The History of a Disease* (Chicago, 1965), viii. According to Veith, the link between witches and hysteria was first made by St. Augustine in the fourth century when he wrote that possession by demons was thought to be evidenced by violent behavioral disorders rather than physical illness. Evil was found not in the body but in the will—placing greater faith in divine intervention than in medicine in alleviating the symptoms. Not until the Enlightenment thinking of the eighteenth century was medical treatment sought for mental afflictions, rather than resorting to religious exorcism. See also 49, 73.

8. Ibid., 235, 245.

9. Michael J. Colligan and Lawrence R. Murphy, "A Review of Mass Psychogenic Illness in Work Settings," in *Mass Psychogenic Illness: A Social Psychological Analysis*, Michael J. Colligan, James W. Pennebaker, and Lawrence R. Murphy, eds. (Hillsdale, N.J., 1982), 47. Also, in the same volume: Joseph E. McGrath, "Complexities, Cautions and Concepts in Research on Mass Psychogenic Illness," 76, and Francois Sirois, "Perspectives on Epidemic Hysteria," 234.

10. Demos, *Entertaining Satan*, 55, 118.

11. Ibid., 127.

12. Marion L. Starkey, *The Devil in Massachusetts: A Modern Enquiry into the Salem Witch Trials* (New York, 1949), 47.

13. Demos, *Entertaining Satan*, 64, 398.

14. Linnda R. Caporeal. "Ergotism: The Satan Loosed in Salem?" *Science* 192 (April 2, 1976), 21–26. The "ergot theory" examines a physiological cause for the problems at Salem.

15. Nicholas P. Spanos and Jack Gottlieb, "Ergotism and the Salem Village Witch Trials," *Science*, vol. 194 (December 24, 1976), 1390–1394. This response to Caporeal's hypothesis noted several reasons why ergotism did not explain the afflictions at Salem and came out immediately after her paper. Nevertheless many people continued to promote the ergot theory for years, possibly because they did not see the refutation by Spanos and Gottlieb or because ergotism seemed so logical.

16. Von Economo, *Encephalitis Lethargica*, 15. Late winter and early spring cases were common in the 1916 epidemic, 12. The skin-surface twitches would have felt like pinches, and an observer could have seen the surface of the skin moving for no reason. Certainly it would have been confusing to both the afflicted and the observer. The capillary bursts may explain the red marks on the girls' arms, which observers likened to marks made by bites or striking with a chain. *Salem Witchcraft Papers*, 402, 599, 736.

17. Zilboorg and Henry, *History*, 210.

18. Ibid., 211.

19. Ibid., 222; *Salem Witchcraft Papers*, 99, 102, 192; Sacks, *Awakenings*, 7; *Salem Witchcraft Papers*, 264, 294.

20. Nicholas Culpeper, "Culpeper's Last Legacy, 1655," in *Health, Medicine, and Mortality in the Sixteenth Century*, Charles Webster, ed. (New York, 1979), 129; R. Pierloot, "Belgium," in Webster, *Health*, 137.

21. Briggs, *Witches & Neighbors*, 51, 55.

22. Ibid., 109.

23. Ibid., 64.

24. Ibid., 68, 73, 74.
25. Ibid., 86.
26. Ibid., 150, 160.
27. Ibid., 282.
28. Ibid., 197.
29. Jonathan Elphick, ed., *Atlas of Bird Migration* (New York, 1995), 86.
30. Ibid., 55.
31. Michael B. A. Oldstone, *Viruses, Plagues, and History* (New York, 1998), 46.
32. Josselyn, *John Josselyn*, 221, 117–119.
33. Thomas W. Scott, "Vertebrate Host Ecology," in Monath, ed., *Arboviruses*, 274.
34. Reiter, "Weather," 250.
35. David W. Stahle, Malcolm K. Cleaveland, Dennis B. Blanton, Matthew D. Therrell, and David A. Gay, "The Lost Colony and Jamestown Droughts," *Science*, vol. 280 (April 24, 1998), 565.
36. Duffy, *Epidemics*, 47. Colonial New England successfully isolated known epidemic infections such as smallpox; in 1648 ships were not allowed to unload passengers if there was any evidence of disease aboard, and communities isolated or restricted ill visitors.
37. Massachusetts Department of Public Health, "Public Health Fact Sheet: Lyme Disease." Available from http://www.state.ma.us/dph/comm; Internet (accessed June 28, 1998).

CHAPTER 7: ALTERNATIVE OUTCOMES

1. William Kidder, *The Diary of Nicholas Gilman* (M.A. thesis, University of New Hampshire, 1972), iv.
2. Ibid., 72, 83, 115; Mrs. M. Grieve, *A Modern Herbal* (New York, 1967), 97.
3. Kidder, *Diary*, 147, 253, 222.
4. Ibid., 242.
5. Ibid., 233, 242.
6. Ibid., 243.
7. Ibid., 255, 256.
8. Ibid., 260, 289.
9. Ibid., 351.
10. Mercy Wheeler, *Address to Young People* (Boston, 1733; reprint, Worcester, Mass.) vi, 9.
11. Ibid., 1, 2.

CHAPTER 8: COULD ENCEPHALITIS LETHARGICA RETURN?

1. Hippocrates, "Sacred," 359.
2. A. Greenough and J. A. Davis, "Encephalitis Lethargica: Mystery of the Past or Undiagnosed Disease of the Present?" *Lancet,* vol. I, no. 8330 (1983), 923.
3. John S. Oxford, "Encephalitis Lethargica: Influenza is a Prime Suspect," *Neurology News,* forthcoming.
4. Cheyette and Cummings, "Encephalitis Lessons," 130–131.
5. Von Economo, *Encephalitis Lethargica,* 167.

CHRONOLOGY

1. Zilboorg and Henry, *History,* 141.
2. Ibid., 210.
3. Josselyn, *John Josselyn,* 89.
4. Demos, *Entertaining Satan,* 346, 373; Duffy, *Epidemics,* 186. Duffy felt it might have been influenza.
5. Demos, *Entertaining Satan,* 373.
6. Duffy, *Epidemics,* 186
7. Ibid., 187.
8. *Salem Witchcraft Papers,* 558.
9. Josselyn, *John Josselyn,* 110; Karlsen, *Devil,* 259.
10. Duffy, *Epidemics,* 187.
11. Demos, *Entertaining Satan,* 373.
12. Knapp's ordeal is detailed in Demos, *Entertaining Satan,* 101; the Boston sickness is in Duffy, *Epidemics,* 203.
13. *Salem Witchcraft Papers,* 124. Carr's symptoms were detailed in court records in 1692, because he felt his illness twenty years earlier (in 1672) was similar to what the afflicted were experiencing.
14. Sacks, *Awakenings,* 13; von Economo, *Encephalitis Lethargica,* 7.
15. Duffy, *Epidemics,* 187; von Economo, *Encephalitis Lethargica,* 8.
16. *Salem Witchcraft Papers,* 92.
17. Ibid., 92–96.
18. Demos, *Entertaining Satan,* 135. These symptoms fit many of the cases cited by Sacks, Tilney, and von Economo. The boy appears to have a classic case of hyperkinetic encephalitis, as described in von Economo, p. 36. The report of his eating sticks may help explain why the afflicted girls were sometimes reported to vomit up pins and other items. Perhaps they too had inexplicably eaten objects during psychosis, then vomited them up later.

19. *Salem Witchcraft Papers*, 496.

20. Ibid., 439.

21. Duffy, *Epidemics*, 207. He thought it was possibly malaria.

22. Ibid., 206; *Salem Witchcraft Papers*, 563. This array of symptoms is nearly exactly like those seen in the young women of Salem eight years later.

23. Duffy, *Epidemics*, 48, 72. *Salem Witchcraft Papers*, 99, 635, 439. The boy had come down with initial symptoms in 1685 and never recovered. His afflictions were included in the 1692 court records in Salem.

24. *Salem Witchcraft Papers*, 95.

25. Ibid., 334.

26. Duffy, *Epidemics*, 188.

27. Ibid. 48.

28. *Salem Witchcraft Papers*, 594.

29. Ibid., 600.

30. Duffy, *Epidemics*, 48; *Salem Witchcraft Papers*, 594, 600, 393.

31. Duffy, *Epidemics*, 48.

32. *Salem Witchcraft Papers*, 94.

33. Duffy, *Epidemics*, 48; *Salem Witchcraft Papers*, 317; Demos, *Entertaining Satan*, 374, 104, 318, 601, 623, 334, 734, 228, 232. This could have been encephalitis, as patients are very often in a coma-like existence for long periods of time. Sacks noted people who had been in similar states for thirty years.

34. Duffy, *Epidemics*, 143.

35. Von Economo, *Encephalitis Lethargica*, 7.

36. Ibid., 188.

37. Ibid., 203.

38. Garrison, *History of Medicine*, 404.

39. Ibid., 117; Kidder, *Diary*, 224; von Economo, *Encephalitis Lethargica*, 7.

40. Von Economo, *Encephalitis Lethargica*, 8.

41. Ibid., 8.

42. Ibid., 8. Also in, "Encephalitis Lethargica," 486.

43. Von Economo, *Encephalitis Lethargica*, 7.

44. "Encephalitis Lethargica," p. 486.

45. Von Economo, *Encephalitis Lethargica*, 11.

BIBLIOGRAPHY

PRIMARY SOURCES

Baganz, Mark D., Peter E. Dross, and John A. Reinhardt. "Rocky Mountain Spotted Fever Encephalitis: MR Findings." *American Journal of Neuroradiology*, vol. 16 (April 1995), 919–922.

Boyer, Paul, and Stephen Nissenbaum, eds. *The Salem Witchcraft Papers*, 3 vols. New York, 1977.

Burr, George Lincoln. *Narratives of the Witchcraft Cases, 1648–1706*. New York, 1914. Reprint, New York, 1968.

Caporeal, Linnda R. "Ergotism: The Satan Loosed in Salem?" *Science*, vol. 192 (April 2, 1976), 21–26. Presents evidence that convulsive ergotism from contaminated grain may have been the physiological basis in 1692 for Salem's witchcraft crisis.

"The Cause of the Epidemic Among Horses in the West." *Science*, vol. 75, no. 1928 (February 19, 1932), 10.

Cheyette, S. R., and J. L. Cummings. "Encephalitis Lethargica: Lessons for Contemporary Neuropsychiatry." *Journal of Neuropsychiatry and Clinical Neurosciences*, vol. 7, no. 2 (Spring 1995), 125–134.

Colligan, Michael J., James W. Pennebaker, and Lawrence R. Murphy, eds. *Mass Psychogenic Illness: A Social Psychological Analysis*. Hillsdale, N.J., 1982. The modern version of mass hysteria theory.

"Disney Takes Precautions Against Encephalitis." (Orlando) *Sun Sentinel*. August 28, 1997.

"Encephalitis Lethargica." *American Journal of Public Health* 13 (June 1923), 486–488.

"Encephalomyelitis." *Science*, supplement, vol. 75, no. 1946 (April 15, 1933), 8.

Jameson, J. Franklin, ed. *Narratives of the Indian Wars, 1675–1699.*
Original Narratives of Early American History series. New
York, 1913.

Josselyn, John, *John Josselyn, Colonial Traveler.* Paul J. Lindholt, ed.
Hanover, N.H., 1988.

Hippocrates. *Writings.* Franklin Center, Pa., 1979.

Kidder, William. "The Diary of Nicholas Gilman." M.A. thesis,
University of New Hampshire, 1972.

Margolis, Simeon, ed., *Johns Hopkins Symptoms and Remedies.* New
York, 1995.

Massachusetts Department of Public Health, Division of Epidemiology and Immunization. "Public Health Fact Sheet:
Lyme Disease." Available from http://www.state.ma.us/dph/
comm; Internet (accessed June 28, 1998).

Mead, Richard. *A Discourse on the Plague.* London, 1744. Reprint,
New York, 1978.

Oxford, John S. "Encephalitis Lethargica: Influenza is a Prime
Suspect." *Neurology News,* forthcoming.

Parris, Samuel. *The Sermon Notebook of Samuel Parris, 1689–1694.*
James F. Cooper, Jr., and Kenneth P. Minkema, eds. Boston,
1993.

Richt, Jurgen A., Isolde Pfeuffer, Mathias Christ, Knut Frese, Karl
Bechter, and Sibylle Herzog. "Borna Disease Virus Infection in
Animals and Humans." *Emerging Infectious Diseases,* vol. 3, no. 3
(July–September 1997). Available from http://www.cdc.gov/
ncidod/EID/vol3no3/richt.htm; Internet (accessed December
24, 1997).

Sacks, Oliver. *Awakenings.* London, 1973. Reprint, New York, 1990.

"Sleeping Sickness." *Science,* supplement (March 30, 1923),
vii–viii.

Stahle, David W., Malcolm K. Cleaveland, Dennis B. Blanton,
Matthew D. Therrell, and David A. Gay. "The Lost Colony
and Jamestown Droughts." *Science,* vol. 280 (April 24, 1998),
564–567.

Tilney, Frederick, and Hubert S. Howe. *Epidemic Encephalitis (Encephalitis Lethargica).* New York, 1920. Includes "Sequelae of
Epidemic Lethargic Encephalitis," reprinted from the *Journal
of the American Medical Association,* vol. 79 (July 15, 1922),
211–214.

Von Economo, Constantin. *Encephalitis Lethargica: Its Sequelae and Treatment.* K. O. Newman, trans. London, 1931.

SECONDARY SOURCES

Adams, James Truslow. *A History of American Life: Provincial Society, 1690–1763,* vol. III. New York, 1927. Useful for its depth and variety of social history.

Alexander, Franz G., and Shelton T. Selesnick. *The History of Psychiatry: An Evaluation of Psychiatric Thought and Practice From Prehistoric Times to the Present.* New York, 1966.

Andrewes, C. H. *The Natural History of Viruses.* New York, 1967.

Bellenir, Karen, and Peter D. Dresser, eds. *Food and Animal Borne Diseases Sourcebook.* Health Reference Series, vol. 7. Detroit, 1995.

Bittle, James L., and Frederick A. Murphy, eds. *Vaccine Biotechnology.* San Diego, 1989.

Boas, Ralph, and Louise Boas. *Cotton Mather, Keeper of the Puritan Conscience.* New York, 1928.

Boorstin, Daniel J. *The Americans: The Colonial Experience.* New York, 1958.

Boyer, Paul, and Stephen Nissenbaum. *Salem Possessed: The Social Origins of Witchcraft.* Cambridge, Mass., 1974.

Briggs, Robin. *Witches & Neighbors: The Social and Cultural Context of European Witchcraft.* New York, 1996.

Calvin, William H. *The Throwing Madonna: Essays on the Brain.* New York, 1983, 1991.

Cartwright, Frederick F. *Disease and History.* New York, 1972.

Creighton, Charles. *A History of Epidemics in Britain, From A.D. 664 to the Extinction of Plague.* Cambridge, England, 1891.

Davidson, James West, and Mark Hamilton Lytle. *After the Fact: The Art of Historical Detection.* New York, 1992.

Demos, John Putnam. *Entertaining Satan: Witchcraft and the Culture of Early New England.* New York, 1982.

———. *The Unredeemed Captive, A Family Story From Early America.* New York, 1994.

Desowitz, Robert S. *Who Gave Pinta to the Santa Maria? Torrid Diseases in a Temperate World.* New York, 1997.

Dourmashkin, R. R. "What Caused the 1918–30 Epidemic of Encephalitis Lethargica?" *Journal of the Royal Society of Medicine,* vol. 90 (1997), 515–520.

Dubos, René. *Man Adapting.* New Haven, Conn., 1965.

Duffy, John. *Epidemics in Colonial America.* Baton Rouge, La., 1953.

Earle, Alice Morse. *Colonial Dames and Good Wives.* Boston, 1895.

———. *Home Life in Colonial Days.* New York, 1931. Reprint of 1898 edition.

Elphick, Jonathan, ed. *Atlas of Bird Migration.* New York, 1995.

Fenner, Frank. *The Biology of Animal Viruses,* vol. II. *The Pathogenesis and Ecology of Viral Infections.* New York, 1968.

Fettner, Ann Giudici. *Viruses: Agents of Change.* New York, 1990.

Garrett, Laurie. *The Coming Plague: Newly Emerging Diseases in a World Out of Balance.* New York, 1994.

Garrison, Fielding H. *The History of Medicine,* 4th ed. Philadelphia, 1913, reprint 1929.

Ginzburg, Carlo. *The Night Battles: Witchcraft and Agrarian Cults in the Sixteenth and Seventeenth Centuries.* Baltimore, 1983.

Greenough, A., and J. A. Davis. "Encephalitis Lethargica: Mystery of the Past or Undiagnosed Disease of the Present?" *Lancet,* vol. I, no. 8330 (1983), 922–923.

Grieve, Mrs. M. *A Modern Herbal.* New York, 1967.

Hall, David D., John M. Murrin, and Thad W. Tate, eds. *Saints and Revolutionaries: Essays on Early American History.* New York, 1984.

Hare, E. H., "Masturbatory Insanity: The History of An Idea." *Journal of Mental Science,* vol. 108 (January 1962), 20.

Hill, Frances. *A Delusion of Satan.* New York, 1995.

Howells, John G., ed. *The World History of Psychiatry.* New York, 1975.

Karlen, Arno. *Man and Microbes: Disease and Plagues in History and Modern Times.* New York, 1995.

Karlsen, Carol F. *The Devil in the Shape of a Woman: Witchcraft in Colonial New England.* New York, 1987.

Kurstak, Edouard, S. J. Lipowski, and P. V. Morozov, eds. *Viruses, Immunity, and Mental Disorders.* New York, 1987.

Louden, Irvine, ed. *Western Medicine: An Illustrated History*. New York, 1997.

Mappen, Marc, ed. *Witches and Historians, Interpretations of Salem*. Malabar, Fla., 1996.

McNeill, William H. *Plagues and Peoples*. New York, 1976. Reprint, New York, 1993.

Monath, Thomas P., ed. *The Arboviruses: Epidemiology and Ecology*, vols. 1–7. Boca Raton, Fla., 1988.

Oldstone, Michael B. A. *Viruses, Plagues, and History*. New York, 1998.

Robinson, Enders A. *The Devil Discovered: Salem Witchcraft 1692*. New York, 1991.

Rosenthal, Bernard. *Salem Story: Reading the Witch Trials of 1692*. Cambridge Studies in American Literature and Culture Series, no. 73. New York, 1993.

Rothschild, Henry, ed. *Biocultural Aspects of Disease*. New York, 1981.

Sagan, Carl. *The Demon Haunted World: Science as a Candle in the Dark*. New York, 1995.

Slotkin, Richard, and James K. Folsom. *So Dreadfull a Judgment: Puritan Responses to King Philip's War, 1676–1677*. Middletown, Conn., 1978.

Spanos, Nicholas P., and Jack Gottlieb. "Ergotism and the Salem Village Witch Trials." *Science*, vol. 194 (December 24, 1976), 1390–1394.

Starkey, Marion L. *The Devil in Massachusetts: A Modern Enquiry Into the Salem Witch Trials*. New York, 1949.

Szasz, Thomas S. *The Manufacture of Madness*. New York, 1970.

Taylor, John M. *The Witchcraft Delusion in Colonial Connecticut, 1647–1697*. New York, 1908.

Tuchman, Barbara W. *Practicing History*. New York, 1982.

Ulrich, Laurel Thatcher. *A Midwife's Tale: The Life of Martha Ballard, Based on Her Diary, 1785–1812*. New York, 1990.

———. *Good Wives: Image and Reality in the Lives of Women in Northern New England, 1650–1750*. New York, 1980.

Van Der Hoeden, J., ed. *Zoonoses*. Amsterdam, 1964.

Veith, Ilza. *Hysteria: The History of a Disease*. Chicago, 1965.

Webster, Charles, ed. *Health, Medicine, and Mortality in the Sixteenth Century*. New York, 1979.

Wertenbaker, Thomas Jefferson. *A History of American Life: The First Americans, 1607–1690*, vol. II. New York, 1927.

Winslow, C. E. A., Wilson G. Smillie, James A. Doull, and John E. Gordon. *The History of American Epidemiology*. St. Louis, 1952.

Zilboorg, Gregory, and George W. Henry. *A History of Medical Psychology*. New York, 1941.

INDEX

A NOTE ON THE AUTHOR

Laurie Winn Carlson is an independent scholar, occasional college teacher, and a writer whose interests have usually centered on the American West. She was born in Sonora, California, and studied at the University of Idaho, Arizona State University, and Eastern Washington University. Her other books include *On Sidesaddles to Heaven: The Women of the Rocky Mountain Mission*; the award-winning children's book *Boss of the Plains: The Hat That Won the West*; and numerous other children's books. She is married with two sons and lives in Cheney, Washington.